Undergraduate Texts in Mathematics

Springer
New York
Berlin
Heidelberg
Barcelona
Hong Kong
London
Milan
Paris
Singapore
Tokyo

Undergraduate Texts in Mathematics

Anglin: Mathematics: A Concise History and Philosophy. *Readings in Mathematics.*
Anglin/Lambek: The Heritage of Thales. *Readings in Mathematics.*
Apostol: Introduction to Analytic Number Theory. Second edition.
Armstrong: Basic Topology.
Armstrong: Groups and Symmetry.
Axler: Linear Algebra Done Right. Second edition.
Beardon: Limits: A New Approach to Real Analysis.
Bak/Newman: Complex Analysis. Second edition.
Banchoff/Wermer: Linear Algebra Through Geometry. Second edition.
Berberian: A First Course in Real Analysis.
Bix: Conics and Cubics: A Concrete Introduction to Algebraic Curves.
Brémaud: An Introduction to Probabilistic Modeling.
Bressoud: Factorization and Primality Testing.
Bressoud: Second Year Calculus. *Readings in Mathematics.*
Brickman: Mathematical Introduction to Linear Programming and Game Theory.
Browder: Mathematical Analysis: An Introduction.
Buskes/van Rooij: Topological Spaces: From Distance to Neighborhood.
Callahan: The Geometry of Spacetime: An Introduction to Special and General Relavitity.
Carter/van Brunt: The Lebesgue–Stieltjes Integral: A Practical Introduction
Cederberg: A Course in Modern Geometries.
Childs: A Concrete Introduction to Higher Algebra. Second edition.
Chung: Elementary Probability Theory with Stochastic Processes. Third edition.
Cox/Little/O'Shea: Ideals, Varieties, and Algorithms. Second edition.
Croom: Basic Concepts of Algebraic Topology.
Curtis: Linear Algebra: An Introductory Approach. Fourth edition.
Devlin: The Joy of Sets: Fundamentals of Contemporary Set Theory. Second edition.
Dixmier: General Topology.
Driver: Why Math?
Ebbinghaus/Flum/Thomas: Mathematical Logic. Second edition.
Edgar: Measure, Topology, and Fractal Geometry.
Elaydi: An Introduction to Difference Equations. Second edition.
Exner: An Accompaniment to Higher Mathematics.
Exner: Inside Calculus.
Fine/Rosenberger: The Fundamental Theory of Algebra.
Fischer: Intermediate Real Analysis.
Flanigan/Kazdan: Calculus Two: Linear and Nonlinear Functions. Second edition.
Fleming: Functions of Several Variables. Second edition.
Foulds: Combinatorial Optimization for Undergraduates.
Foulds: Optimization Techniques: An Introduction.
Franklin: Methods of Mathematical Economics.
Frazier: An Introduction to Wavelets Through Linear Algebra.
Gordon: Discrete Probability.
Hairer/Wanner: Analysis by Its History. *Readings in Mathematics.*
Halmos: Finite-Dimensional Vector Spaces. Second edition.
Halmos: Naive Set Theory.
Hämmerlin/Hoffmann: Numerical Mathematics. *Readings in Mathematics.*
Harris/Hirst/Mossinghoff: Combinatorics and Graph Theory.
Hartshorne: Geometry: Euclid and Beyond.
Hijab: Introduction to Calculus and Classical Analysis.

(continued after index)

M. Carter B. van Brunt

The Lebesgue-Stieltjes Integral

A Practical Introduction

With 45 Illustrations

Springer

M. Carter
B. van Brunt
Institute of Fundamental Sciences
Palmerston North Campus
Private Bag 11222
Massey University
Palmerston North 5301
New Zealand

Mathematics Subject Classification (2000): 28-01

Library of Congress Cataloging-in-Publication Data
Carter, M. (Michael), 1940–
The Lebesgue–Stieltjes integral : a practical introduction / M. Carter, B. van Brunt.
p. cm. – (Undergraduate texts in mathematics)
Includes bibliographical references and index.
ISBN 0-387-95012-5 (alk. paper)
1. Lebesgue integral. I. van Brunt, B. (Bruce) II. Title. III. Series.
QA312.C37 2000
515′.43–dc21 00-020065

Printed on acid-free paper.

Production managed by Timothy Taylor; manufacturing supervised by Jerome Basma.
Typeset by The Bartlett Press Inc., Marietta, GA.
Printed and bound by R.R. Donnelley and Sons, Harrisonburg, VA.
Printed in the United States of America.

9 8 7 6 5 4 3 2 1

ISBN 0-387-95012-5 Springer-Verlag New York Berlin Heidelberg SPIN 10756530

Preface

It is safe to say that for every student of calculus the first encounter with integration involves the idea of approximating an area by summing rectangular strips, then using some kind of limit process to obtain the exact area required. Later the details are made more precise, and the formal theory of the Riemann integral is introduced.

The budding pure mathematician will in due course top this off with a course on measure and integration, discovering in the process that the Riemann integral, natural though it is, has been superseded by the Lebesgue integral and other more recent theories of integration. However, those whose interests lie more in the direction of applied mathematics will in all probability find themselves needing to use the Lebesgue or Lebesgue–Stieltjes integral without having the necessary theoretical background. Those who try to fill this gap by doing some reading are all too often put off by having to plough through many pages of preliminary measure theory.

It is to such readers that this book is addressed. Our aim is to introduce the Lebesgue–Stieltjes integral on the real line in a natural way as an extension of the Riemann integral. We have tried to make the treatment as practical as possible. The evaluation of Lebesgue–Stieltjes integrals is discussed in detail, as are the key theorems of integral calculus such as integration by parts and change of variable, as well as the standard convergence theorems. Multivariate integrals are discussed briefly, and practical results such as Fubini's theorem are highlighted. The final chapters of the book are devoted to the Lebesgue integral and its role in analysis. Specifically, function spaces based on the Lebesgue integral are discussed along with some elementary results.

While we have developed the theory rigorously, we have not striven for completeness. Where a rigorous proof would require lengthy preparation, we have not hesitated to state important theorems without proof in order to keep the book reasonably brief and accessible. There are many excellent treatises on integration that provide complete treatments for those who are interested.

The book could also be used as a textbook for a course on integration for nonspecialists. Indeed, it began life as a set of notes for just such a course. We have included a number of exercises that extend and illustrate the theory and provide practice in the techniques. Hints and answers to these problems are given at the end of the book.

We have assumed that the reader has a reasonable knowledge of calculus techniques and some acquaintance with basic real analysis. The early chapters deal with the additional specialized concepts from analysis that we need. The later chapters discuss results from functional analysis. It is intended that these chapters be essentially self-contained; no attempt is made to be comprehensive, and numerous references are given for specific results.

Michael Carter
Bruce van Brunt
Palmerston North, New Zealand

Contents

CHAPTER 1 Real Numbers

The field of mathematics known as analysis, of which integration is a part, is characterized by the frequent appeal to limiting processes. The properties of real numbers play a fundamental role in analysis. Indeed, it is through a limiting process that the real number system is formally constructed. It is beyond the scope of this book to recount this construction. We shall, however, discuss some of the properties of real numbers that are of immediate importance to the material that will follow in later chapters.

1.1 Rational and Irrational Numbers

The number systems of importance in real analysis include the natural numbers ($\mathbb{N}$), the integers ($\mathbb{Z}$), the rational numbers ($\mathbb{Q}$), and the real numbers ($\mathbb{R}$). The reader is assumed to have some familiarity with these number systems. In this section we highlight some of the properties of the rational and irrational numbers that will be used later.

The set of real numbers can be partitioned into the subsets of rational and irrational numbers. Recall that rational numbers are numbers that can be expressed in the form m/n, where m and n are integers with $n \neq 0$ (for example $\frac{3}{4}$, $\frac{11}{2}$, $-\frac{5}{7}(=\frac{-5}{7})$, $15(=\frac{15}{1})$, $0(=\frac{0}{1})$). Irrational numbers are characterized by the property that they cannot be expressed as the quotient of two integers. Numbers such as e, π, and $\sqrt{2}$ are familiar examples of irrational numbers.

It follows at once from the ordinary arithmetic of fractions that if r_1 and r_2 are rational numbers, then so are $r_1 + r_2$, $r_1 - r_2$, $r_1 r_2$, and r_1/r_2 (in the last case, provided that $r_2 \neq 0$). Using these facts we can prove the following theorem:

Theorem 1.1.1

If r is a rational number and x is an irrational number, then

(i) $r + x$ is irrational;

(ii) rx is irrational, provided that $r \neq 0$.

Proof See Exercises 1-1, No. 1. □

A fundamental property of irrational and rational numbers is that they are both "dense" on the real line. The precise meaning of this is given by the following theorem:

Theorem 1.1.2

If a and b are real numbers with $a < b$, then there exist both a rational number and an irrational number between a and b.

Proof Let a and b be real numbers such that $a < b$. Then $b - a > 0$, so $\sqrt{2}/(b-a) > 0$. Let k be an integer less than a, and let n be an integer such that $n > \sqrt{2}/(b-a)$. Then

$$0 < \frac{1}{n} < \frac{\sqrt{2}}{n} < b - a,$$

and so the succesive terms of each of the sequences

$$k + \frac{1}{n},\quad k + \frac{2}{n},\quad k + \frac{3}{n}, \ldots$$

$$k + \frac{\sqrt{2}}{n},\quad k + 2\frac{\sqrt{2}}{n},\quad k + 3\frac{\sqrt{2}}{n}, \ldots$$

differ by less than the distance between a and b. Thus at least one term of each sequence must lie beween a and b. But the terms of the first sequence are all rational, while (by Theorem 1.1.1) those of the second are all irrational, so the theorem is proved. □

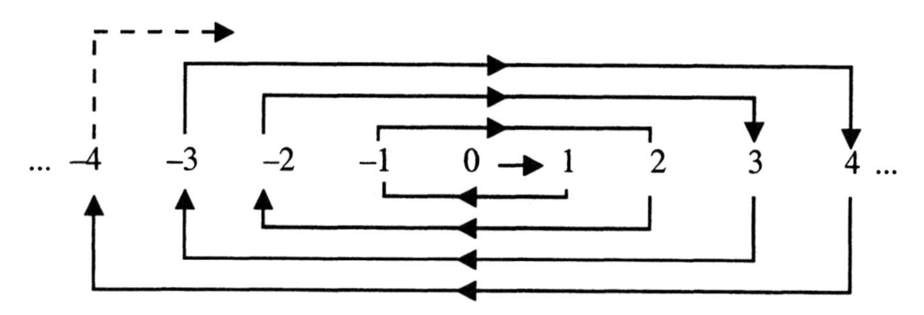

FIGURE 1.1 Counting the integers

Corollary 1.1.3
If a and b are real numbers with $a < b$, then between a and b there exist infinitely many rational numbers and infinitely many irrational numbers.

Proof This follows immediately by repeated application of Theorem 1.1.2. □

An infinite set S is said to be **countable** if there is a one-to-one correspondence between the elements of S and the natural numbers. In other words, S is countable if its elements can be listed as a sequence

$$S = \{a_1, a_2, a_3, \ldots\}.$$

For example, the set $\mathbb{Z}$ is countable because its elements can be listed as a sequence $\{a_1, a_2, a_3, \ldots\}$ by using the rule

$$a_n = \begin{cases} 0 & \text{if } n = 1 \\ m & \text{if } n = 2m,\ m > 0 \\ -m & \text{if } n = 2m + 1,\ m > 0 \end{cases}$$

so that $a_1 = 0$, $a_2 = 1$, $a_3 = -1$, $a_4 = 2$, and so on. The process of listing the elements of $\mathbb{Z}$ as a sequence can be visualized by following the arrows in Figure 1.1 starting at 0. Much less obvious is the fact that the set $\mathbb{Q}$ is also countable. Figure 1.2 depicts a scheme for counting the rationals. To list the rationals as a sequence we can just follow the arrowed path in Figure 1.2 starting at $0/1 = 0$, and omitting any rational number that has already been listed. The set

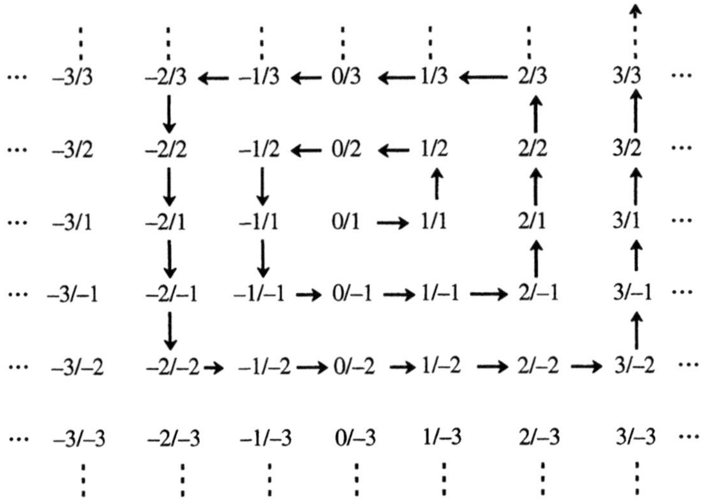

	⋮	⋮	⋮	⋮	⋮	⋮	⋮
⋯	−3/3	−2/3	−1/3	0/3	1/3	2/3	3/3 ⋯
⋯	−3/2	−2/2	−1/2	0/2	1/2	2/2	3/2 ⋯
⋯	−3/1	−2/1	−1/1	0/1	1/1	2/1	3/1 ⋯
⋯	−3/−1	−2/−1	−1/−1	0/−1	1/−1	2/−1	3/−1 ⋯
⋯	−3/−2	−2/−2	−1/−2	0/−2	1/−2	2/−2	3/−2 ⋯
⋯	−3/−3	−2/−3	−1/−3	0/−3	1/−3	2/−3	3/−3 ⋯
	⋮	⋮	⋮	⋮	⋮	⋮	⋮

FIGURE 1.2 Counting the rationals

$\mathbb{Q}$ can thus be written as

$$\mathbb{Q} = \left\{0, 1, \frac{1}{2}, -\frac{1}{2}, -1, -2, 2, \frac{2}{3}, \frac{1}{3}, -\frac{1}{3}, -\frac{2}{3}, -\frac{3}{2}, -3, 3, \ldots\right\}.$$

The infinite sets $\mathbb{N}$, $\mathbb{Z}$, and $\mathbb{Q}$ are all countable, and one may wonder whether in fact there are any infinite sets that are *not* countable. The next theorem settles that question:

Theorem 1.1.4
The set S of all real numbers x such that $0 \leq x < 1$ is not countable.

Proof We use without proof here the well-known fact that any real number can be represented in decimal form. This representation is not unique, because $N.n_1n_2n_3\ldots n_k9999\ldots$ and $N.n_1n_2n_3\ldots(n_k + 1)0000\ldots$ are the same number (e.g. $2.349999\ldots = 2.35$); likewise $N.999\ldots$ and $N + 1$ are the same number. We can make the representation unique by choosing the second of these representations in all such cases, so that none of our decimal expressions will end with recurring 9's.

We will use a proof by contradiction to establish the theorem. Suppose S is countable, so that we can list all the elements of S as a sequence:

$$S = \{a_1, a_2, a_3, \ldots\}.$$

Now, each element of this sequence can be represented in decimal form, say

$$a_n = 0.x_{n1}x_{n2}x_{n3}x_{n4}\ldots,$$

where for all $n, j \in \mathbb{N}$, x_{nj} is one of the digits $0, 1, 2, \ldots, 9$. The elements of S can thus be written in the form

$$\begin{aligned}
a_1 &= 0.x_{11}x_{12}x_{13}x_{14}\ldots,\\
a_2 &= 0.x_{21}x_{22}x_{23}x_{24}\ldots,\\
a_3 &= 0.x_{31}x_{32}x_{33}x_{34}\ldots,\\
a_4 &= 0.x_{41}x_{42}x_{43}x_{44}\ldots,\\
&\vdots
\end{aligned}$$

We define a real number $b = 0.m_1m_2m_3m_4\ldots$, where for each $j \in \mathbb{N}$,

$$m_j = \begin{cases} 1 & \text{if } x_{jj} \neq 1,\\ 2 & \text{if } x_{jj} = 1. \end{cases}$$

Suppose, for example, that our listing of elements of S begins

$$\begin{aligned}
a_1 &= 0.\underline{8}37124\ldots,\\
a_2 &= 0.1\underline{1}2563\ldots,\\
a_3 &= 0.33\underline{3}333\ldots,\\
a_4 &= 0.258\underline{6}14\ldots,\\
&\vdots
\end{aligned}$$

Then:

$$\begin{aligned}
x_{11} &= 8 \neq 1 && \text{so } m_1 = 1,\\
x_{22} &= 1 && \text{so } m_2 = 2,\\
x_{33} &= 3 \neq 1 && \text{so } m_3 = 1,\\
x_{44} &= 6 \neq 1 && \text{so } m_4 = 1,
\end{aligned}$$

and so on. The decimal expansion of b therefore begins $0.1211\ldots$. It is clear that $0 < b < 1$, so that $b \in S$, and therefore we must have $b = a_N$ for some $N \in \mathbb{N}$. But by definition, the decimal expansion of b differs from that of a_N at the Nth decimal place, so $b \neq a_N$ and we have a contradiction. We thus conclude that our original assumption must be false, and S cannot be countable. □

It follows at once from this theorem that the set $\mathbb{R}$ is not countable. In fact, it is also not hard to deduce that the set of all real numbers belonging to any interval of nonzero length (however small) is not countable.

Exercises 1-1:

1. Use the method of proof by contradiction to prove Theorem 1.1.1.
2. Give examples to show that if x_1 and x_2 are irrational numbers, then $x_1 + x_2$ and $x_1 x_2$ may be rational or irrational.
3. Since the set of all rational numbers is countable, it follows easily that the set $S^* = \{x : 0 \leq x < 1 \text{ and } x \text{ rational}\}$ is countable. Thus, if we apply the argument used in the proof of Theorem 1.1.4 to S^* instead of S, something must go wrong with the argument. What goes wrong?
4. (a) Prove that the union of two countable sets is countable.
   (b) Use a proof by contradiction to prove that the set of all irrational numbers is not countable.

1.2 The Extended Real Number System

It is convenient to introduce at this point a notation that is useful in many parts of analysis; care, however, should be taken not to read too much into it.

The **extended real number system** is defined to be the set $\mathbb{R}_e$ consisting of all the real numbers together with the symbols ∞ and $-\infty$, in which the operations of addition, subtraction, multiplication, and division between real numbers are as in the real number system, and the symbols ∞ and $-\infty$ have the following properties for any $x \in \mathbb{R}$:

(i) $-\infty < x < \infty$;

(ii) $\infty + x = x + \infty = \infty$ and $-\infty + x = x + (-\infty) = -\infty$;

(iii) $\infty + \infty = \infty$ and $-\infty + (-\infty) = -\infty$;

(iv) $\infty \cdot x = x \cdot \infty = \infty$ and $(-\infty) \cdot x = x \cdot (-\infty) = -\infty$ for any $x > 0$;

(v) $\infty \cdot x = x \cdot \infty = -\infty$ and $(-\infty) \cdot x = x \cdot (-\infty) = \infty$ for any $x < 0$;

(vi) $\infty \cdot \infty = \infty$, $\infty \cdot (-\infty) = (-\infty) \cdot \infty = -\infty$, and $(-\infty) \cdot (-\infty) = \infty$.

The reader is warned that the new symbols ∞ and $-\infty$ are defined only in terms of the above properties and cannot be used except as prescribed by these conventions. In particular, expressions such as $\infty + (-\infty)$, $(-\infty) + \infty$, $\infty \cdot 0$, $0 \cdot \infty$, $0 \cdot (-\infty)$, and $(-\infty) \cdot 0$ are *meaningless*.

A number $a \in \mathbb{R}_e$ is said to be **finite** if $a \in \mathbb{R}$, i.e. if a is an ordinary real number.

In all that follows, when we say that I is an *interval with endpoints* a, b we mean that a and b are elements of $\mathbb{R}_e$ (unless specifically restricted to finite values) with $a \leq b$, and I is one of the following subsets of $\mathbb{R}$:

(i) the **open** interval $\{x \in \mathbb{R} : a < x < b\}$, denoted by (a, b);

(ii) the **closed** interval $\{x \in \mathbb{R} : a \leq x \leq b\}$, denoted by $[a, b]$, where a and b must be finite;

(iii) the **closed-open** interval $\{x \in \mathbb{R} : a \leq x < b\}$, denoted by $[a, b)$, where a must be finite;

(iv) the **open-closed** interval $\{x \in \mathbb{R} : a < x \leq b\}$, denoted by $(a, b]$, where b must be finite.

Note that although the endpoints of an interval may not be finite, the actual elements of the interval are finite. Note also that for any $a \in \mathbb{R}$, the set $[a, a]$ consists of the single point a, whereas the sets $[a, a)$ and $(a, a]$ are both empty. The interval (a, a) is empty for all $a \in \mathbb{R}_e$.

The only change from standard interval notation is that intervals such as $(-\infty, -3]$, $(-\infty, \infty)$, $(-2, \infty)$, etc. are defined. (Intervals such as $[-\infty, 3]$, $[-\infty, \infty]$, $(-2, \infty]$, etc. are not.)

1.3 Bounds

Let S be any nonempty subset of $\mathbb{R}_e$. A number $c \in \mathbb{R}_e$ is called an **upper bound** of S if $x \leq c$ for all $x \in S$. Similarly, a number $d \in \mathbb{R}_e$ is called a **lower bound** of S if $x \geq d$ for all $x \in S$.

Evidently, ∞ is an upper bound and $-\infty$ is a lower bound for any nonempty subset of $\mathbb{R}_e$. In general, most subsets will have many upper and lower bounds. For example, consider the set $S_1 = (-3, 2]$. Any number $c \in \mathbb{R}_e$ such that $c \geq 2$ is an upper bound of S_1, and any number $d \in \mathbb{R}_e$ such that $d \leq -3$ is a lower bound of S_1. Note that there is a least upper bound for S_1 (namely 2) and that in fact it is also an element of S_1. Note also that there is a greatest lower bound (namely -3), which is not a member of S_1.

As another example, consider the set

$$S_2 = \left\{\frac{1}{n} : n \in \mathbb{N}\right\} = \left\{1, \frac{1}{2}, \frac{1}{3}, \ldots\right\}.$$

Here any $c \geq 1$ is an upper bound of S_2, while any $d \leq 0$ is a lower bound. Note that no positive number can be a lower bound of S_2, because for any $d > 0$ we can always find a positive integer n sufficiently large so that $1/n < d$, and therefore d cannot be a lower bound of S_2. Thus S_2 has a least upper bound 1 and a greatest lower bound 0.

As a final example, let $S_3 = \mathbb{Q}$. Then ∞ is the only upper bound of S_3 and $-\infty$ is the only lower bound. Thus S_3 has a least upper bound ∞ and a greatest lower bound $-\infty$.

The following result (often taken as an axiom), which we state without proof, expresses a fundamental property of the extended real number system:

Theorem 1.3.1
Any nonempty subset of $\mathbb{R}_e$ has both a least upper bound and a greatest lower bound in $\mathbb{R}_e$.

The least upper bound of a nonempty set $S \subseteq \mathbb{R}_e$ is often called the **supremum** of S and is denoted by $\sup S$; the greatest lower bound of S is often called the **infimum** of S and denoted by $\inf S$. The examples given above indicate that $\sup S$ and $\inf S$ may or may not be elements of S; however, in the case where $\sup S$ or $\inf S$ is *finite*, although $\sup S$ and $\inf S$ need not be in S, they must at any rate be "close" to S in a sense that is made precise by the following theorem:

Theorem 1.3.2

Let $S \subseteq \mathbb{R}_e$ be nonempty.

(i) *If $M \in \mathbb{R}_e$ is* finite, *then $M = \sup S$ if and only if M is an upper bound of S and for each real number $\epsilon > 0$ (however small) there exists a number $x \in S$ (depending on ϵ) such that $M - \epsilon < x \leq M$.*

(ii) *If m is* finite, *then $m = \inf S$ if and only if m is a lower bound of S and for each real number $\epsilon > 0$ (however small) there exists a number $x \in S$ (depending on ϵ) such that $m \leq x < m + \epsilon$.*

Proof We shall prove part (i) and leave part (ii) as an exercise. Suppose $M = \sup S$, where M is finite. Then M, being the least upper bound of S, is certainly an upper bound of S. Let ϵ be any positive real number. Then $M - \epsilon < M$, and so $M - \epsilon$ cannot be an upper bound of S, since M is the least upper bound. Thus there must exist a number $x \in S$ such that $x > M - \epsilon$, and since we know that M is an upper bound of S, we have $M - \epsilon < x \leq M$.

Conversely, suppose that M is finite, M is an upper bound of S, and that for any real number $\epsilon > 0$ there exists a number $x \in S$ such that $M - \epsilon < x \leq M$. Let K be any finite element of $\mathbb{R}_e$ with $K < M$. Then $M - K > 0$, so taking $\epsilon = M - K$ we have that there exists an $x \in S$ such that $M - (M - K) < x \leq M$, i.e., $K < x \leq M$. Thus K cannot be an upper bound of S, and since $-\infty$ is obviously not an upper bound of S, it follows that M must be the least upper bound of S. □

Exercises 1-3:

1. Give the least upper and greatest lower bounds of each of the following subsets of $\mathbb{R}_e$, and state in each case whether or not they are elements of the set in question:

(a) $\{x : 0 \leq x \leq 5\}$
(b) $\{x : 0 < x \leq 5\}$
(c) $\{x : x^2 > 3\}$
(d) $\{x : \frac{1}{x} > 2\}$
(e) $\{x : x \text{ is rational and } x^2 < 2\}$
(f) $\{x : x = 3 + \frac{1}{n^2},\ n \in \mathbb{N}\}$
(g) $\{x : x \text{ is rational and positive}\}$

2. If $S \subseteq \mathbb{R}_e$ has only finitely many elements, say $S = \{x_1, x_2, \ldots, x_n\}$, then clearly S has both a greatest element and a least element, denoted by $\max\{x_1, x_2, \ldots, x_n\}$ and $\min\{x_1, x_2, \ldots, x_n\}$, respectively. Prove:

$$\sup\{x_1, x_2, \ldots, x_n\} = \max\{x_1, x_2, \ldots, x_n\},$$
$$\inf\{x_1, x_2, \ldots, x_n\} = \min\{x_1, x_2, \ldots, x_n\}.$$

3. Prove that if S_1 and S_2 are nonempty subsets of $\mathbb{R}_e$ such that $S_1 \subseteq S_2$, then $\sup S_1 \leq \sup S_2$ and $\inf S_1 \geq \inf S_2$.

4. Let S be a nonempty subset of $\mathbb{R}_e$, and c a nonzero real number. Define S^* by $S^* = \{cx : x \in S\}$.

   (a) Prove that if c is positive, then $\sup S^* = c(\sup S)$ and $\inf(S^*) = c(\inf S)$.

   (b) Prove that if c is negative, then $\sup S^* = c(\inf S)$ and $\inf(S^*) = c(\sup S)$.

5. Prove part (ii) of Theorem 1.3.2.

2 CHAPTER Some Analytic Preliminaries

Before we can develop the theory of integration, we need to revisit the concept of a sequence and deal with a number of topics in analysis involving sequences, series, and functions.

2.1 Monotone Sequences

Convergence of a sequence on $\mathbb{R}_e$ can be defined in a manner analogous to the usual definition for sequences on $\mathbb{R}$. Specifically, a sequence $\{a_n\}$ on $\mathbb{R}_e$ is said to **converge** to a finite limit if there is a finite number $a \in \mathbb{R}_e$ having the property that given any positive real number ϵ (however small) there is a number $N \in \mathbb{N}$ such that $|a_n - a| < \epsilon$ whenever $n \geq N$. This relationship is expressed by $a_n \to a$ as $n \to \infty$, or simply $a_n \to a$. The number a is called the **limit** of the sequence.

If for any finite number $M \in \mathbb{R}_e$ there exists an $N \in \mathbb{N}$ such that $a_n > M$ whenever $n \geq N$, then we write $a_n \to \infty$ as $n \to \infty$ or simply $a_n \to \infty$, and the limit of the sequence is said to be ∞; similarly, if for any finite number $M \in \mathbb{R}_e$ there exists an $N \in \mathbb{N}$ such that $a_n < M$ whenever $n \geq N$, then we write $a_n \to -\infty$ as $n \to \infty$ or simply $a_n \to -\infty$, and the limit of the sequence is said to be $-\infty$.

Let $\{a_n\}$ be a sequence of real numbers. The sequence $\{a_n\}$ is said to be **monotone increasing** if $a_n \leq a_{n+1}$ for all $n \in \mathbb{N}$, and **monotone decreasing** if $a_n \geq a_{n+1}$ for all $n \in \mathbb{N}$. For example:

The sequence $1, 2, 3, 4, \ldots$ is monotone increasing.
The sequence $1, \frac{1}{2}, \frac{1}{3}, \frac{1}{4}, \ldots$ is monotone decreasing.
The sequence $1, 1, 2, 2, 3, 3, \ldots$ is monotone increasing.
The sequence $1, 1, 1, 1, \ldots$ is monotone increasing and monotone decreasing.
The sequence $1, 0, 1, 0, \ldots$ is neither monotone increasing nor monotone decreasing.

If a sequence $\{a_n\}$ is monotone increasing with limit $\ell \in \mathbb{R}_e$, we write $a_n \uparrow \ell$ (read "a_n increases to ℓ"). If the sequence is monotone decreasing with limit $\ell \in \mathbb{R}_e$, we write $a_n \downarrow \ell$ (read "a_n decreases to ℓ").

We shall frequently be studying sequences of functions. Let $\{f_n\}$ denote a sequence of functions $f_n : I \to \mathbb{R}$ defined on some interval $I \subseteq \mathbb{R}$. The sequence $\{f_n\}$ is said to converge on I to a function f if for each $x \in I$ the sequence $\{f_n(x)\}$ converges to $f(x)$, i.e., if the sequence is pointwise convergent. The notation used for sequences of functions is similar to that used for sequences of numbers: specifically,

$f_n \to f$ on I means that for each $x \in I$, $f_n(x) \to f(x)$.
$f_n \uparrow f$ on I means that for each $x \in I$, $f_n(x) \uparrow f(x)$.
$f_n \downarrow f$ on I means that for each $x \in I$, $f_n(x) \downarrow f(x)$.

The fundamental theorem concerning monotone sequences is the following:

Theorem 2.1.1

Let $\{a_n\}$ be a sequence on $\mathbb{R}$.

(i) If the sequence $\{a_n\}$ is monotone increasing, then $a_n \uparrow \sup\{a_n\}$.
(ii) If the sequence $\{a_n\}$ is monotone decreasing, then $a_n \downarrow \inf\{a_n\}$.

Proof We shall prove part (i) of the theorem, leaving the second part as an exercise. Let $M = \sup\{a_n\}$. The proof of part (i) can be partitioned into two cases depending on whether or not M is finite.

Case 1: If $M = \infty$, then for any positive real number K, we know that K cannot be an upper bound of $\{a_n\}$, so there exists a positive integer N such that $a_N > K$. Since the sequence is monotone increasing, it follows that $a_n \geq a_N > K$ for all $n \geq N$, and thus $a_n \uparrow \infty (= M)$ by definition.

Case 2: Suppose M finite and let ϵ be any positive real number. Then by Theorem 1.3.2 there exists a positive integer N such that

$$M - \epsilon < a_N \leq M.$$

Since the sequence is monotone increasing and has M as an upper bound, it follows that

$$M - \epsilon < a_N \leq a_n \leq M < M + \epsilon$$

for all $n \geq N$. This implies that for all $n \geq N$,

$$|a_n - M| < \epsilon$$

and consequently $a_n \to M$ by definition. Since the sequence is monotone increasing, this means that $a_n \uparrow M$ as required. □

Exercises 2-1:

1. Let S be a nonempty subset of $\mathbb{R}$, with $\sup S = M$ and $\inf S = m$. Show that there exist sequences $\{a_n\}$ and $\{b_n\}$ of elements of S such that $a_n \uparrow M$ and $b_n \downarrow m$.
2. Prove part (ii) of Theorem 2.1.1.

2.2 Double Series

Let $\{a_n\}$ be a sequence on $\mathbb{R}_e$. Recall that the infinite series $\sum_{m=1}^{\infty} a_m$ is said to **converge** if the sequence of partial sums $\{s_n\}$, where $s_n = \sum_{m=1}^{n} a_m$, converges to a finite number. If $s_n \to \infty$, then the series is said to diverge to ∞; if $s_n \to -\infty$, then the series is said to diverge to $-\infty$. Often, questions concerning the convergence of an infinite

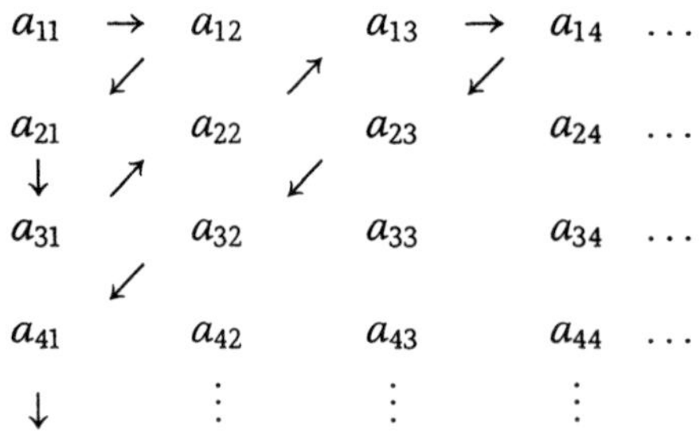

$$
\begin{array}{cccccccc}
a_{11} & \rightarrow & a_{12} & & a_{13} & \rightarrow & a_{14} & \cdots \\
 & \swarrow & & \nearrow & & \swarrow & & \\
a_{21} & & a_{22} & & a_{23} & & a_{24} & \cdots \\
\downarrow & \nearrow & & \swarrow & & & & \\
a_{31} & & a_{32} & & a_{33} & & a_{34} & \cdots \\
 & \swarrow & & & & & & \\
a_{41} & & a_{42} & & a_{43} & & a_{44} & \cdots \\
\downarrow & & \vdots & & \vdots & & \vdots &
\end{array}
$$

FIGURE 2.1

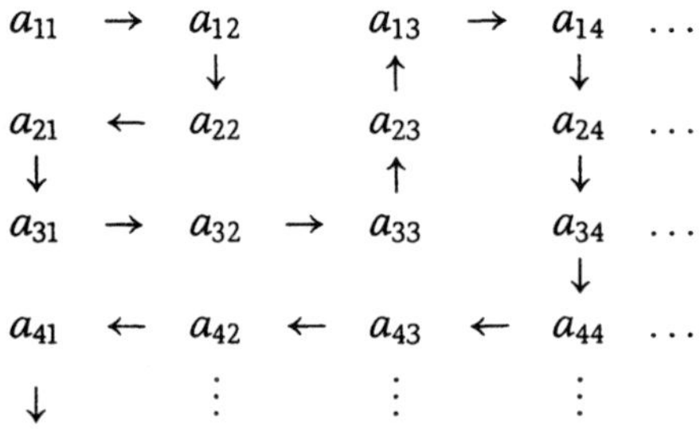

$$
\begin{array}{cccccccc}
a_{11} & \rightarrow & a_{12} & & a_{13} & \rightarrow & a_{14} & \cdots \\
 & & \downarrow & & \uparrow & & \downarrow & \\
a_{21} & \leftarrow & a_{22} & & a_{23} & & a_{24} & \cdots \\
\downarrow & & & & \uparrow & & \downarrow & \\
a_{31} & \rightarrow & a_{32} & \rightarrow & a_{33} & & a_{34} & \cdots \\
 & & & & & & \downarrow & \\
a_{41} & \leftarrow & a_{42} & \leftarrow & a_{43} & \leftarrow & a_{44} & \cdots \\
\downarrow & & \vdots & & \vdots & & \vdots &
\end{array}
$$

FIGURE 2.2

series involve considering sequences $\{a_n\}$ of nonnegative terms (e.g., absolute convergence). If the terms of the sequence $\{a_n\}$ consist of nonnegative numbers, then the resulting sequence of partial sums is monotone increasing. Theorem 2.1.1 thus implies that $s_n \uparrow \sup\{s_n\}$ and therefore that either the series $\sum_{m=1}^{\infty} a_m$ converges or it diverges to ∞, according as $\sup\{s_n\}$ is finite or ∞.

Consider the array of real numbers depicted in Figure 2.1. This array can be written as a (single) sequence in many ways. One way is to follow the arrowed path in the diagram. This gives the sequence

$$\{a_{11}, a_{12}, a_{21}, a_{31}, a_{22}, a_{13}, a_{14}, a_{23}, \ldots\},$$

but this is obviously not the only way. Another scheme for constructing a sequence is given in Figure 2.2.

For any way of writing this array as a single sequence A_1, A_2, $A_3, \ldots$ we can form the corresponding infinite series $\sum_{j=1}^{\infty} A_j$. We know from Riemann's theorem on the derangement of series [6] that in general, the convergence and limit of the series depends on the particular sequence $\{A_n\}$ used, but there are some situations in which *every* possible sequence leads to the same answer. When this is the case, it is sensible to introduce the notion of a "double series" $\sum_{m,n=1}^{\infty} a_{mn}$ and consider questions such as convergence. This leads us to the following definition: If for *all possible* ways of writing the array $\{a_{mn}\}$ as a single sequence the corresponding series has the finite sum ℓ, then the *double series* $\sum_{m,n=1}^{\infty} a_{mn}$ is said to **converge** to ℓ. If for all possible ways of writing the array as a single sequence the corresponding series either always diverges to ∞ or always diverges to $-\infty$, then the double series is said to be **properly divergent** (to ∞ or $-\infty$ as the case may be). In all other circumstances the double series is simply said to be divergent, and its sum does not exist as an element of $\mathbb{R}_e$.

As well as "summing" the array by writing it as a single sequence, we can "sum" it by first summing the rows and then adding the sums of the rows, giving the **repeated series** $\sum_{m=1}^{\infty}(\sum_{n=1}^{\infty} a_{mn})$. Alternatively, we can first sum the columns and then add the sums of the columns, giving the repeated series $\sum_{n=1}^{\infty}(\sum_{m=1}^{\infty} a_{mn})$.

The relationship between convergence for a double series $\sum_{m,n=1}^{\infty} a_{mn}$ and for the two related repeated series is, in general, complicated. For our purposes, however, we can focus on the particularly simple case where all of the entries in the array are nonnegative, i.e., $a_{mn} \geq 0$ for all $n, m \in \mathbb{N}$. In this case we have the following result, which is stated without proof:

Theorem 2.2.1

Suppose that for all $n, m \in \mathbb{N}$ we have $a_{mn} \geq 0$, where $a_{mn} \in \mathbb{R}_e$. Then the double series $\sum_{m,n=1}^{\infty} a_{mn}$ and the two repeated series $\sum_{n=1}(\sum_{m=1} a_{mn})$ and $\sum_{m=1}(\sum_{n=1} a_{mn})$ either all converge to the same finite sum or are all properly divergent to ∞.

More details on double series can be found in [6].

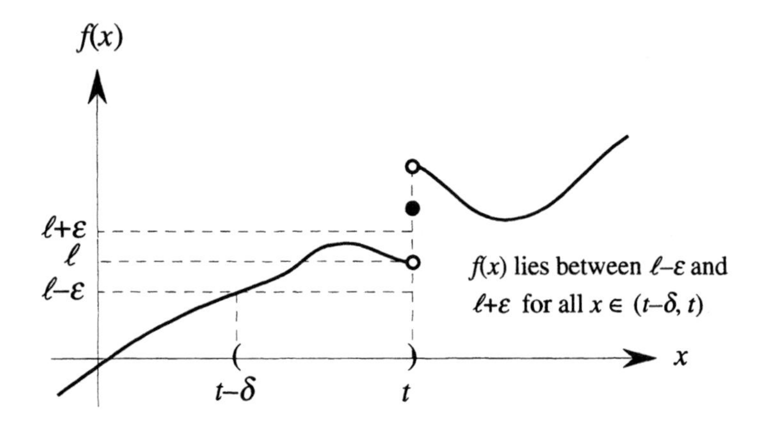

FIGURE 2.3

2.3 One-Sided Limits

Let $f : \mathbb{R} \to \mathbb{R}$ be a function, and t and ℓ real numbers. Recall that $\lim_{x\to t} f(x) = \ell$ if and only if for any positive real number ϵ, however small, there exists a positive real number δ such that

$$0 < |x - t| < \delta \Longrightarrow |f(x) - \ell| < \epsilon.$$

We say that $\lim_{x\to t} f(x) = \infty$ if for any number M there exists a $\delta > 0$ such that $f(x) > M$ whenever $0 < |x - t| < \delta$. A similar definition can be made for $\lim_{x\to t} f(x) = -\infty$. In these definitions x can be either to the left or the right of t, i.e., x is free to approach t from the left or right (or for that matter oscillate on either side of t). Often it is of use to restrict the manner in which x approaches t, particularly if no information about f is available on one side of t, or t lies at the end of the interval under consideration. For these situations it is useful to introduce the notion of limits from the left and from the right. Such limits are referred to as **one-sided limits**.

The **limit from the left** is defined as follows: $\lim_{x\to t^-} f(x) = \ell$ if and only if for any positive real number ϵ there exists a positive real number δ such that

$$t - \delta < x < t \Longrightarrow |f(x) - \ell| < \epsilon$$

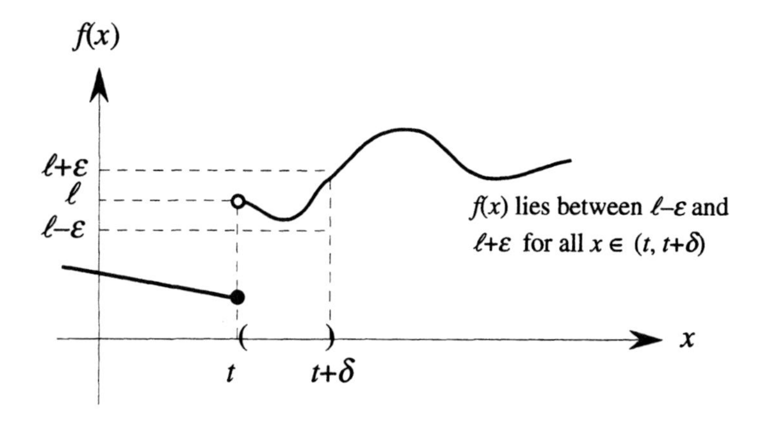

FIGURE 2.4

(cf. Figure 2.3). In this case we say that $f(x)$ tends to ℓ as x tends to t from the left. Similarly, the **limit from the right** is defined as $\lim_{x\to t^+} f(x) = \ell$ if and only if for any positive real number ϵ there exists a positive real number δ such that

$$t < x < t + \delta \Longrightarrow |f(x) - \ell| < \epsilon$$

(cf. Figure 2.4). In this case we say that $f(x)$ tends to ℓ as x tends to t from the right.

We can easily extend these definitions for cases where the limit is not finite, e.g., $\lim_{x\to t^-} f(x) = \infty$ if and only if for any positive real number M there exists a positive real number δ such that

$$t - \delta < x < t \Longrightarrow f(x) > M.$$

Example 2-3-1:
Let $f : \mathbb{R} \to \mathbb{R}$ be defined as

$$f(x) = \begin{cases} -1 & \text{if } x < 1, \\ 0 & \text{if } x = 1, \\ x/2 & \text{if } x > 1. \end{cases}$$

Then $\lim_{x\to 1^-} f(x) = -1$ and $\lim_{x\to 1^+} f(x) = 1/2$. This function is depicted in Figure 2.5.

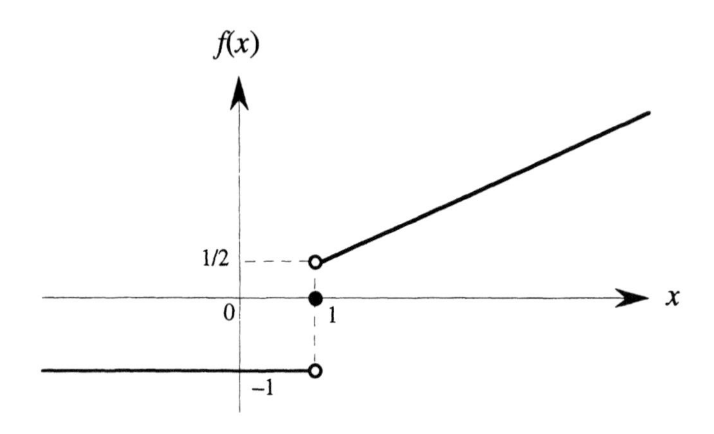

FIGURE 2.5

Example 2-3-2:
Let $f(x) = 1/(x-1)$ (cf. Figure 2.6). Then $\lim_{x\to 1^-} f(x) = -\infty$ and $\lim_{x\to 1^+} f(x) = \infty$.

The definition of a limit can be extended further to consider cases where $x \to \infty$ or $x \to -\infty$. For example, let $a \in \mathbb{R}$. Then $\lim_{x\to\infty} f(x) = a$ if and only if for any positive real number ϵ there exists a number X such that

$$x > X \Longrightarrow |f(x) - a| < \epsilon.$$

Definitions similar to the finite case can also be framed for $\lim_{x\to-\infty} f(x) = a$, $\lim_{x\to\infty} f(x) = \infty$, and $\lim_{x\to-\infty} f(x) = \infty$, etc. The usual elementary rules for limits of sums, differences, products, and quotients of functions hold for one-sided limits just as for ordinary limits. For example, if $\lim_{x\to t^-} f(x) = a$ and $\lim_{x\to t^-} g(x) = b$, then $\lim_{x\to t^-}(f(x)+g(x)) = a+b$, $\lim_{x\to t^-}(f(x)g(x)) = ab$, etc. These relations are proved the same way as for the ordinary limit case. It is also easy to prove that $\lim_{x\to t} f(x) = \ell$ if and only if $\lim_{x\to t^-} f(x) = \ell$ and $\lim_{x\to t^+} f(x) = \ell$.

For succinctness, we shall often denote $\lim_{x\to t^-} f(x)$ by $f(t^-)$ and $\lim_{x\to t^+} f(x)$ by $f(t^+)$. In some circumstances we will denote $\lim_{x\to\infty} f(x)$ by $f(\infty^-)$ and $\lim_{x\to-\infty} f(x)$ by $f((-\infty)^+)$.

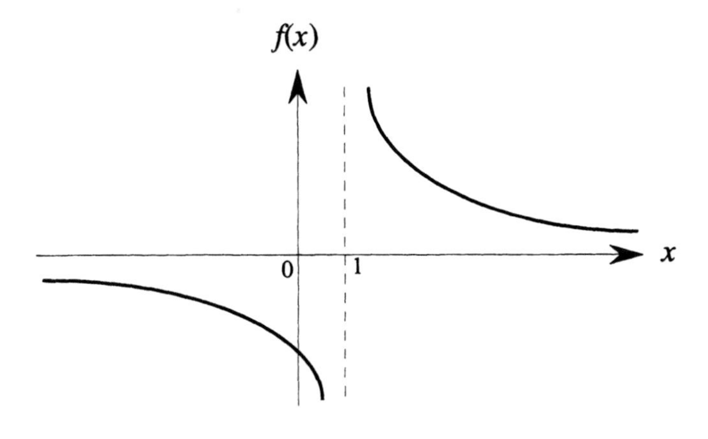

FIGURE 2.6

One-sided continuity for a function f at finite points t is defined in terms of one-sided limits in the obvious way. We say that f is **continuous on the left** at t if $f(t)$ is defined and finite, $f(t^-)$ exists, and $f(t^-) = f(t)$, and **continuous on the right** at t if $f(t)$ is defined and finite, $f(t^+)$ exists, and $f(t^+) = f(t)$. Evidently, f is continuous at t if and only if it is both continuous on the left and continuous on the right at t, i.e., if and only if $f(t^-) = f(t) = f(t^+)$.

There are several different ways in which a function can fail to be continuous at a point. If $f(t^-), f(t), f(t^+)$ all exist but are not all equal, then f is said to have a **jump discontinuity** at t. Thus, the function in Example 2-3-1 has a jump discontinuity at 1. A function may fail to be continuous at a point because the limit is not finite. The function of Example 2-3-2 is discontinuous at 1 not only because the limit is not finite but also because $f(1^-) \neq f(1^+)$ and $f(t)$ has not been defined. Yet another way in which a function can fail to be continuous at a point is when the right or left limits fail to exist. The next example illustrates this.

Example 2-3-3:
Consider the function $f : \mathbb{R} \to \mathbb{R}$ defined by

$$f(x) = \begin{cases} \sin(1/x), & \text{if } x \neq 0, \\ 0, & \text{if } x = 0. \end{cases}$$

Figure 2.7 illustrates this function. Now, $|\sin(1/x)| \leq 1$ and

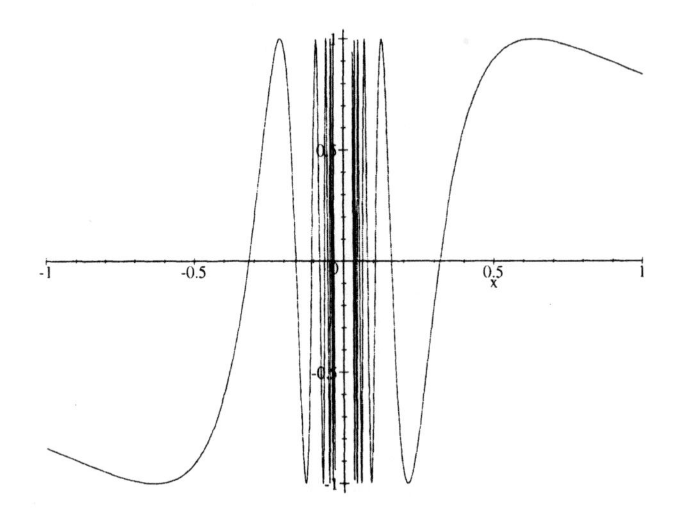

FIGURE 2.7

$\sin(1/x) = 0$ if and only if $1/x = n\pi$, where $n \in \mathbb{Z} - \{0\}$, i.e., when $x = 1/(n\pi)$. Moreover, $\sin(1/x) = 1$ if and only if $1/x = (4n+1)\pi/2$, where $n \in \mathbb{Z}$, i.e., $x = 2/((4n+1)\pi)$, and $\sin(1/x) = -1$ if and only if $1/x = (4n+3)\pi/2$, where $n \in \mathbb{Z}$, i.e., $x = 2/((4n+3)\pi)$. Near $x = 0$, x attains the values $1/(n\pi)$, $2/((4n+1)\pi)$, $2/((4n+3)\pi)$ infinitely many times (for different $n \in \mathbb{Z}$), and thus it can be shown that neither $f(0^-)$ nor $f(0^+)$ exists, so f is discontinuous at 0. The function oscillates infinitely often in any interval $(-\delta, \delta)$, $\delta > 0$.

2.4 Monotone Functions

Let $f : \mathbb{R} \to \mathbb{R}$ be a function. We say that f is **monotone increasing** if $f(x_1) \leq f(x_2)$ whenever $x_1 < x_2$. The function f is said to be **monotone decreasing** if $f(x_1) \geq f(x_2)$ whenever $x_1 < x_2$. If f is either monotone increasing or monotone decreasing, then it is said to be **monotone**. Some examples are:

(i) The function in Example 2-3-1 is monotone increasing.

(ii) The function $|x|$ is neither monotone increasing nor monotone decreasing.

(iii) Constant functions are both monotone increasing and monotone decreasing.

One can also speak of functions being monotone increasing or monotone decreasing on a particular interval rather than the entire real line. For example, the function $|x|$ is monotone decreasing on $(-\infty, 0]$ and monotone increasing on the interval $[0, \infty)$. In this section, however, we will restrict the discussion to functions that are monotone on the entire real line. The general case will be discussed in Section 2.7.

The most important theorem on monotone functions is the following:

Theorem 2.4.1
Let $f : \mathbb{R} \to \mathbb{R}$ be a monotone function. Then, for all $t \in \mathbb{R}$, $f(t^-)$ and $f(t^+)$ exist and are finite, and also $f(\infty^-)$ and $f((-\infty)^+)$ exist, but are not necessarily finite. Furthermore, for all $t \in \mathbb{R}$,
(i) if f is monotone increasing, then $f(t^-) \le f(t) \le f(t^+)$;
(ii) if f is monotone decreasing, then $f(t^-) \ge f(t) \ge f(t^+)$.

Proof Suppose f is monotone increasing, and let t be any real number. Let $m = \inf\{f(x) : t < x\}$ and $M = \sup\{f(x) : x < t\}$. Now, $f(t)$ is finite, and since f is monotone increasing, $f(t)$ is a lower bound of $\{f(x) : t < x\}$ and an upper bound of $\{f(x) : x < t\}$. It follows that m and M are finite, and also

$$M \le f(t) \le m. \tag{2.1}$$

Now take any $\epsilon > 0$. By Theorem 1.3.2, there exist x_1 and x_2, with $t < x_1$ and $t > x_2$, such that $m \le f(x_1) < m + \epsilon$ and $M - \epsilon < f(x_2) \le M$. Since f is monotone increasing and m is a lower bound of $\{f(x) : t < x\}$, it follows that

$$t < x < x_1 \Longrightarrow m \le f(x) \le f(x_1) < m + \epsilon \Longrightarrow |f(x) - m| < \epsilon$$

and similarly

$$x_2 < x < t \Longrightarrow M - \epsilon < f(x_2) \le f(x) \le M \Longrightarrow |f(x) - M| < \epsilon.$$

Thus, by definition, $f(t^+) = m$ and $f(t^-) = M$. Also, statement (i) follows from equation (2.1).

Next, let $A = \inf\{f(x) : x \in \mathbb{R}\}$; here, A may be finite, or equal to $-\infty$. If A is finite, an argument similar to that used previously shows that $f((-\infty)^+) = A$. If A is $-\infty$, let K be any negative real number. Then K is not a lower bound of $\{f(x) : x \in \mathbb{R}\}$, so there exists an $x_1 \in \mathbb{R}$ such that $f(x_1) < K$. Since f is monotone increasing, it follows that

$$x < x_1 \Longrightarrow f(x) \leq f(x_1) < K,$$

and so $f((-\infty)^+) = -\infty = A$ in this case also. A similar argument shows that $f(\infty^-) = \sup\{f(x) : x \in \mathbb{R}\}$.

The case where f is monotone decreasing can be proved in a similar way, or by considering the function $-f$ (see Exercises 2-4, No. 1). □

Corollary 2.4.2

(i) *If f is monotone increasing, and a, b are elements of $\mathbb{R}_e$ with $a < b$, then $f(a^+) \leq f(b^-)$.*

(ii) *If f is monotone decreasing, and a, b are elements of $\mathbb{R}_e$ with $a < b$, then $f(a^+) \geq f(b^-)$.*

Proof We will prove part (i) of this theorem and leave the other part as an exercise. Let f be monotone increasing. From the proof of Theorem 2.4.1 we know that $f(a^+) = \inf\{f(x) : a < x\}$ and $f(b^-) = \sup\{f(x) : x < b\}$. Since $a < b$, there exists a $y \in \mathbb{R}$ such that $a < y < b$, and so $f(a^+) \leq f(y)$ and $f(y) \leq f(b^-)$, whence $f(a^+) \leq f(b^-)$ as required. □

If f is monotone, then for any real t we have by Theorem 2.4.1 that $f(t^-), f(t)$, and $f(t^+)$ all exist. It follows at once that the only discontinuities that a monotone function can have are jump discontinuities.

In general, a function $f : \mathbb{R} \to \mathbb{R}$ may have any number of points of discontinuity. Indeed, the function f defined by

$$f(x) = \begin{cases} 0, & \text{if } x \text{ is rational,} \\ 1, & \text{if } x \text{ is irrational,} \end{cases}$$

is discontinuous at every real number. However, for monotone functions we have the following theorem:

Theorem 2.4.3
If $f : \mathbb{R} \to \mathbb{R}$ is monotone, then the set of points at which f is discontinuous is either empty, finite, or countably infinite.

Proof If f is monotone decreasing, then $-f$ is monotone increasing (see Exercises 2-4, No. 1)) and has the same points of discontinuity as f, so it is sufficient to prove the theorem for the case where f is monotone increasing.

Let E be the set of points at which f is discontinuous, and suppose E is not empty. Then for each $x \in E$ we have $f(x^-) < f(x^+)$, and so by Theorem 1.1.2 there exists a rational number r_x such that $f(x^-) < r_x < f(x^+)$. Now by Corollary 2.4.2 we have $x_1 < x_2 \Longrightarrow f(x_1^+) \leq f(x_2^-)$, and it follows that if $x_1, x_2 \in E$ are such that $x_1 < x_2$, then $r_{x_1} < r_{x_2}$; thus, we have associated with each $x \in E$ a distinct rational number.

Since the set of all rational numbers can be listed as a sequence, it follows that the set $\{r_x : x \in E\}$ can also be listed as a (finite or infinite) sequence. We can then list the elements of E in the same order as their associated rational numbers. Thus E (if not empty) is either finite or countably infinite. □

Although Theorem 2.4.3 places restrictions on the possible set of discontinuities of a monotone function, this set can nevertheless be quite complicated, and one must be careful not to make unjustified assumptions about it. For example, one might guess that the discontinuities of a monotone function must be some minimum distance apart, but the following example shows that this need not be so.

Example 2-4-1:
Let $f : \mathbb{R} \to \mathbb{R}$ be defined as follows:

$$f(x) = \begin{cases} 0, & \text{if } x \leq 0, \\ 1/(n+1), & \text{if } 1/(n+1) < x \leq 1/n,\ n = 1, 2, 3, \ldots, \\ 1, & \text{if } x > 1. \end{cases}$$

Figure 2.8 illustrates this function. Clearly, f is monotone increasing. It can be shown that $f(0^+) = 0$ (see Exercises 2-4, No. 3), so f has jump discontinuities at the countably infinite set of points $\{1, \frac{1}{2}, \frac{1}{3}, \frac{1}{4}, \ldots\}$ and is continuous at all other points. In fact, unlikely

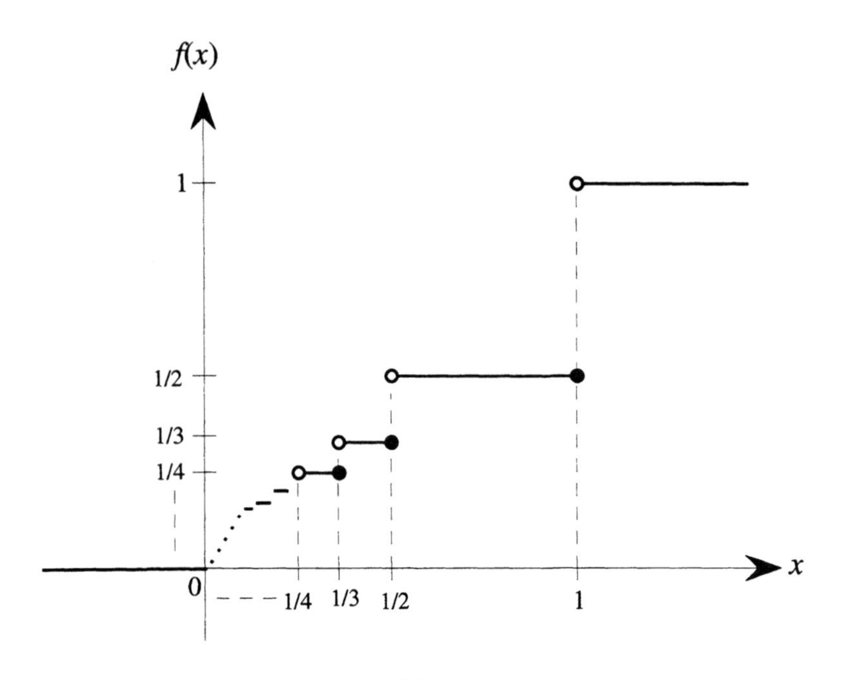

FIGURE 2.8

as it may seem, it is possible to construct a monotone increasing function that is discontinuous at every rational number!

Exercises 2-4:

1. Prove part (ii) of Theorem 2.4.1, by showing that if f is monotone decreasing, then $-f$ is monotone increasing, and then applying part (i).
2. Prove part (ii) of Corollary 2.4.2.
3. Prove that $f(0^+) = 0$ in Example 2-4-1.

2.5 Step Functions

Let I be any interval. A function $\theta : I \to \mathbb{R}$ is called a **step function** if there is a finite collection $\{I_1, I_2, \cdots, I_n\}$ of pairwise disjoint intervals such that $S = I_1 \cup I_2 \cup \cdots \cup I_n \subseteq I$ and a set $\{c_1, c_2, \ldots, c_n\}$

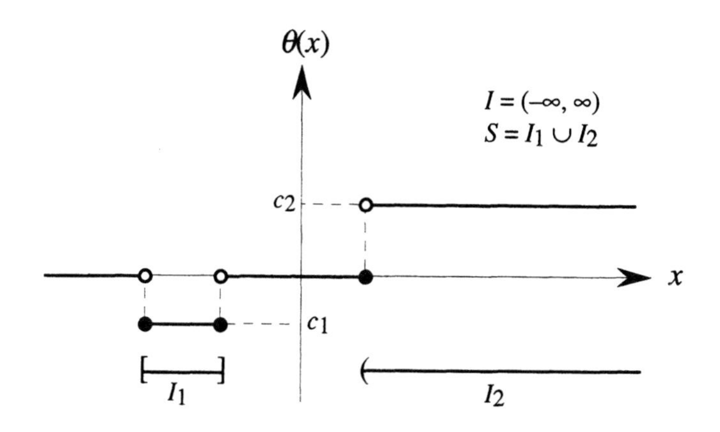

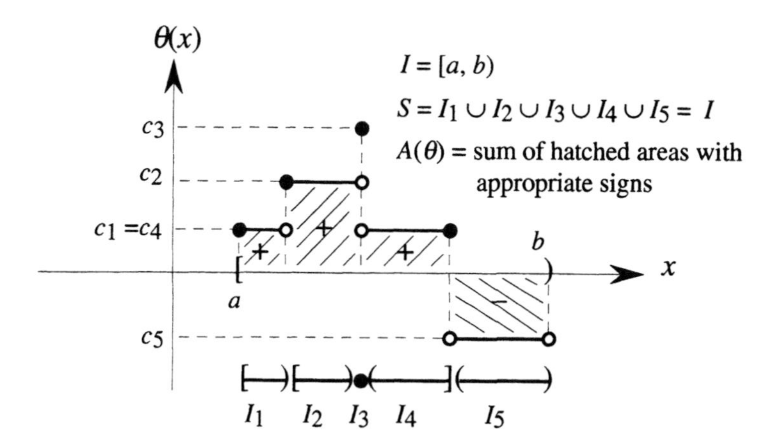

FIGURE 2.9

of finite, nonzero real numbers such that

$$\theta(x) = \begin{cases} c_j, & \text{if } x \in I_j, j = 1, 2, \ldots, n, \\ 0, & \text{if } x \in I - S. \end{cases}$$

In other words, θ is constant and nonzero on each interval I_j, and zero elsewhere in I. The set S on which θ is nonzero is called the **support** of θ. Note that S may be empty, so that the zero function on I is also a step function. Figure 2.9 illustrates some possible step-function configurations.

If the support of a step function θ has finite total length, then we associate with θ the area $A(\theta)$ between the graph of θ and the x-axis, with the usual convention that areas below the x-axis have negative sign (we often refer to $A(\theta)$ as the "area under the graph" of θ). Thus $A(\theta)$ exists for the step function θ in Figure 2.9-2, but not for that in Figure 2.9-1.

If $\theta_1, \theta_2, \ldots, \theta_m$ are step functions on the same interval I, all with supports of finite total length, and if $a_1, a_2, \ldots, a_m$ are finite real numbers, then the function θ defined by

$$\theta(x) = \sum_{j=1}^{m} a_j \theta_j(x)$$

for $x \in I$ is also a step function on I. The support of θ has finite length, and

$$A(\theta) = \sum_{j=1}^{m} a_j A(\theta_j).$$

The fact that θ is also a step function is a rather tedious and messy thing to prove in detail, but an example should be sufficient to indicate why it is true.

Example 2-5-1:
Let $\theta_1, \theta_2 : [0, 3) \to \mathbb{R}$ be defined by

$$\theta_1(x) = \begin{cases} 1, & \text{if } 0 \le x < 2, \\ 2, & \text{if } 2 \le x < 3, \end{cases} \qquad \theta_2(x) = \begin{cases} -1, & \text{if } 0 \le x \le 1, \\ 1, & \text{if } 1 < x < 3 \end{cases}$$

(cf. Figure 2.10). Let $\theta = 2\theta_1 - \theta_2$. Then

$$\theta(x) = \begin{cases} 3, & \text{if } 0 \le x \le 1, \\ 1, & \text{if } 1 < x < 2, \\ 3, & \text{if } 2 \le x < 3. \end{cases}$$

(cf. Figure 2.11). Clearly, θ is a step function. Note also that

$$A(\theta_1) = 2(1) + 1(2) = 4,$$
$$A(\theta_2) = -1(1) + 2(1) = 1,$$

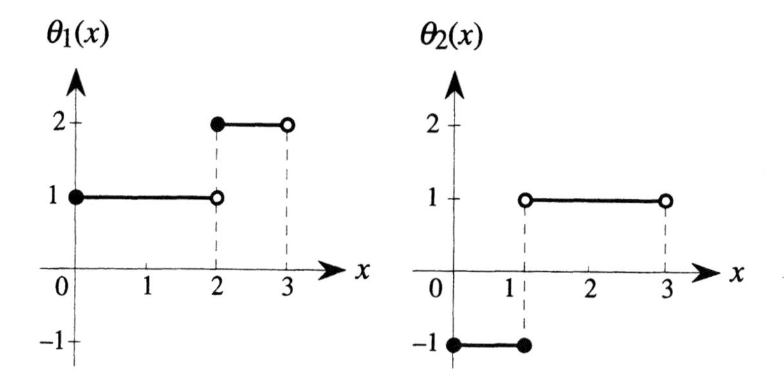

FIGURE 2.10

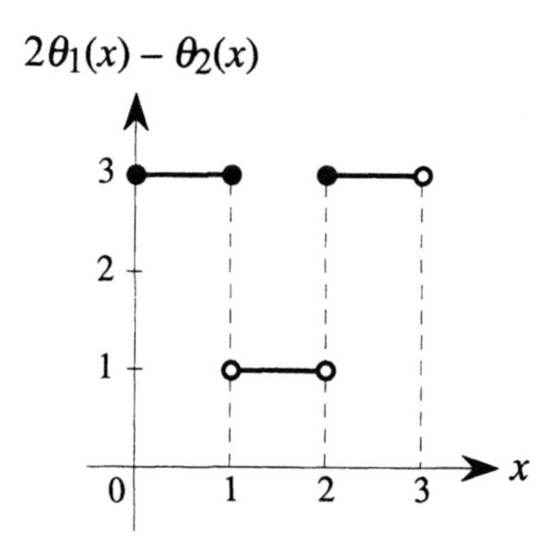

FIGURE 2.11

$$
\begin{aligned}
A(\theta) &= 1(3) + 1(1) + 1(3) = 7 \\
&= 2A(\theta_1) - A(\theta_2),
\end{aligned}
$$

as expected.

If $f, g : I \to \mathbb{R}$ are such that $f(x) \leq g(x)$ for all $x \in I$, we write simply "$f \leq g$ on I." The following properties of areas under graphs of step functions are geometrically obvious and straightforward to prove:

(i) If $\theta \geq 0$ on I, and the support of θ has finite total length, then $A(\theta) \geq 0$. Also, $A(0) = 0$.

(ii) If θ_1 and θ_2 both have supports of finite total length, and $\theta_1 \leq \theta_2$ on I, then $A(\theta_1) \leq A(\theta_2)$.

Exercises 2-5:

1. Let $\theta_1, \theta_2 : [0, 3] \to \mathbb{R}$ be defined by

$$\theta_1(x) = \begin{cases} 1, & \text{if } 0 \leq x < 1, \\ -1, & \text{if } 1 \leq x \leq 2, \\ 4, & \text{if } 2 < x \leq 3, \end{cases} \qquad \theta_2(x) = \begin{cases} -2, & \text{if } 0 \leq x \leq 1, \\ 3, & \text{if } 1 < x \leq 3. \end{cases}$$

Sketch the graphs of θ_1, θ_2, and $\theta_1 - 2\theta_2$, and verify by direct calculation that $A(\theta_1 - 2\theta_2) = A(\theta_1) - 2A(\theta_2)$.

2. Let $\theta_1, \theta_2 : \mathbb{R} \to \mathbb{R}$ be defined by

$$\theta_1(x) = \begin{cases} 0, & \text{if } x \leq -1, \\ 1, & \text{if } -1 < x \leq 2, \\ 0, & \text{if } x > 2, \end{cases} \qquad \theta_2(x) = \begin{cases} 0, & \text{if } x \leq 0, \\ -1, & \text{if } 0 < x < 3, \\ 0, & \text{if } x \geq 3. \end{cases}$$

Sketch the graphs of θ_1, θ_2, and $\theta_1 + \theta_2$, and verify by direct calculation that $A(\theta_1 + \theta_2) = A(\theta_1) + A(\theta_2)$.

2.6 Positive and Negative Parts of a Function

Let I be any interval. For any function $f : I \to \mathbb{R}$ we define the functions $f^+ : I \to \mathbb{R}$ and $f^- : I \to \mathbb{R}$, called the **positive part** and the **negative part** of f, respectively, as follows:

$$f^+(x) = \max\{f(x), 0\} \text{ for all } x \in I,$$
$$f^-(x) = \min\{f(x), 0\} \text{ for all } x \in I.$$

We also define the function $|f| : I \to \mathbb{R}$ by

$$|f|(x) = |f(x)|,$$

for all $x \in I$. These definitions are depicted graphically in Figure 2.12. It is clear that for any function $f : I \to \mathbb{R}$, we have $f = f^+ + f^-$

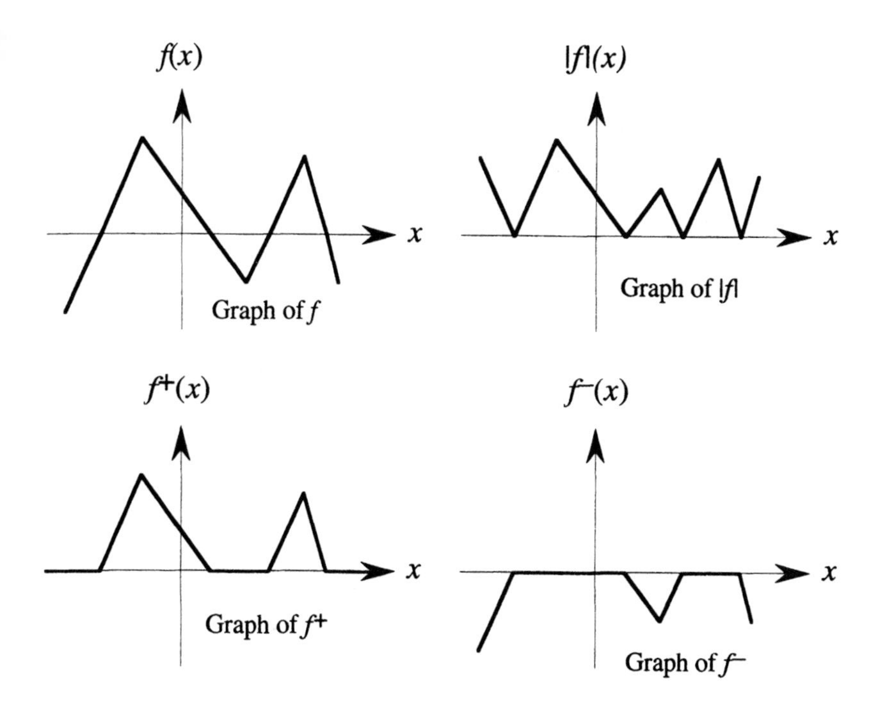

FIGURE 2.12

and $|f| = f^+ - f^-$. It is also clear that $0 \leq f^+ \leq |f|$ and $-|f| \leq f^- \leq 0$ on I.

Exercises 2-6: If $f, g : I \to \mathbb{R}$, prove the following inequalities:

1. $|f^+ - g^+| \leq |f - g|$ on I.
2. $|f^- - g^-| \leq |f - g|$ on I.
3. $||f| - |g|| \leq |f - g|$ on I.

2.7 Bounded Variation and Absolute Continuity

For any (nonempty) interval I, a **partial subdivision** of I is a collection $S = \{I_1, I_2, \ldots, I_n\}$ of *closed* intervals such that:

(i) $I_1 \cup I_2 \cup \cdots \cup I_n \subseteq I$;

(ii) for any $j, k = 1, 2, \ldots, n$ with $j \neq k$, either $I_k \cap I_j$ is empty or $I_k \cap I_j$ consists of a single point that is an endpoint of both I_j and I_k.

For example, if $I = [0, 3)$, then $S = \{[0, 1], [1, \frac{3}{2}], [2, \frac{5}{2}]\}$ is a partial subdivision of I.

Let $f : I \to \mathbb{R}$ be a function, and let $S = \{I_1, I_2, \ldots, I_n\}$ be a partial subdivision of I. For each $j = 1, 2, \ldots, n$, let I_j have endpoints a_j, b_j. We can associate with f, I, and S the quantity $V_S(f, I)$ defined by

$$V_S(f, I) = \sum_{j=1}^{n} |f(b_j) - f(a_j)|.$$

Consider now the set $A(f, I) = \{V_S(f, I)\colon S \text{ is a partial subdivision of } I\}$. Obviously $V_S(f, I)$ cannot be negative, so 0 is a lower bound of $A(f, I)$. The least upper bound of $A(f, I)$ is called the **total variation** of f over I, and denoted by $V(f, I)$; and we have $0 \leq V(f, I) \leq \infty$ for any f and I.

Example 2-7-1:
Let $f : I \to \mathbb{R}$ be any step function. If f is constant on I, then evidently $V(f, I) = 0$. If not, then as x increases through I, $f(x)$ has a finite number of changes in value. Let the absolute magnitudes of these changes be $k_1, k_2, \ldots, k_m$.

Now take any closed interval $I_j = [a_j, b_j] \subseteq I$. If none of the changes in the value of $f(x)$ occur within I_j, then $f(x)$ is constant on I_j and $|f(b_j) - f(a_j)| = 0$. If the changes numbered $r_1, r_2, \ldots, r_p$ occur within I_j, then $|f(b_j) - f(a_j)| \leq \sum_{i=1}^{p} k_{r_i}$.

If S is a partial subdivision of I, then since a given change in the value of $f(x)$ can occur within at most one of the intervals $I_1, I_2, \ldots, I_n$ that make up S, it follows that $V_S(f, I) \leq \sum_{r=1}^{m} k_r$. Furthermore, if we choose S such that each change in the value of $f(x)$ occurs within one of the intervals comprising S, and no interval has more than one change occurring within it, then $V_S(f, I) = \sum_{r=1}^{m} k_r$. It follows that f has finite total variation given by

$$V(f, I) = \sum_{r=1}^{m} k_r,$$

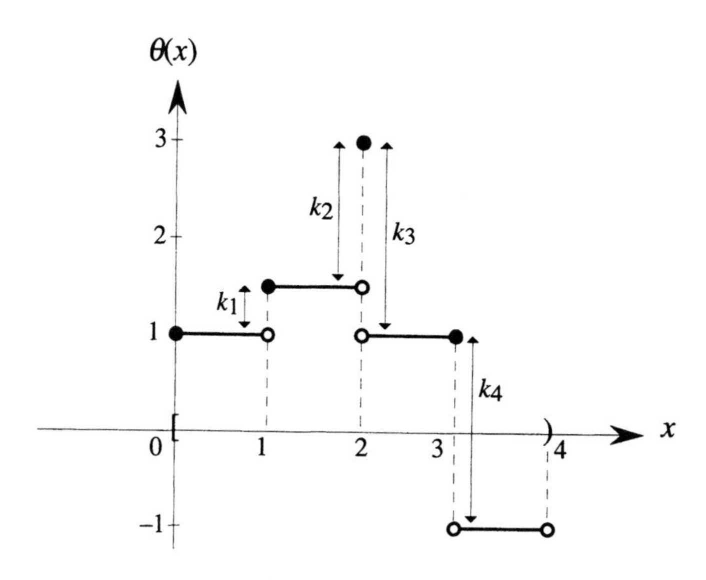

FIGURE 2.13

where $\sum_{r=1}^{m} k_r$ is the sum of absolute values of all changes in the value of $f(x)$.

For instance, let $\theta : [0, 4) \to \mathbb{R}$ be defined by

$$\theta_1(x) = \begin{cases} 1, & \text{if } 0 \leq x < 1, \\ \frac{3}{2}, & \text{if } 1 \leq x < 2, \\ 3, & \text{if } x = 2, \\ 1, & \text{if } 2 < x \leq 3, \\ -1, & \text{if } 3 < x < 4. \end{cases}$$

(cf. Figure 2.13). Numbering the changes in value of $\theta(x)$ from left to right, their absolute magnitudes are $k_1 = \frac{1}{2}$, $k_2 = \frac{3}{2}$, $k_3 = 2$, $k_4 = 2$, respectively, and the sum of the absolute magnitudes of all the changes is therefore $\sum_{r=1}^{4} k_r = 6$.

Let $S = \{[\frac{1}{2}, \frac{3}{2}], [\frac{3}{2}, 2], [2, \frac{5}{2}], [3, \frac{7}{2}]\}$. Then

$$\begin{aligned} V_S(\theta, [0, 4)) = &|\theta(\frac{3}{2}) - \theta(\frac{1}{2})| + |\theta(2) - \theta(\frac{3}{2})| \\ &+ |\theta(\frac{5}{2}) - \theta(2)| + |\theta(\frac{7}{2}) - \theta(3)| \end{aligned}$$

$$= |\frac{3}{2} - 1| + |3 - \frac{3}{2}| + |1 - 3| + |-1 - 1|$$
$$= \frac{1}{2} + \frac{3}{2} + 2 + 2,$$

and so $V_S(\theta, [0, 4)) = \sum_{r=1}^{4} k_r$. Note that exactly one of the four changes in the value of $\theta(x)$ occurs within each of the four intervals making up S.

Example 2-7-2:
Let $f : (0, 1) \to \mathbb{R}$ be defined by $f(x) = \sin(1/x)$ for all $x \in (0, 1)$. (The graph of this function is depicted in Figure 2.7.) For each $j = 1, 2, \ldots, n$, let

$$I_j = \left[\frac{2}{(j+1)\pi}, \frac{2}{j\pi}\right].$$

Then $S = \{I_1, I_2, \ldots, I_n\}$ is a partial subdivision of $(0, 1)$, and we have

$$V_S(f, (0, 1)) = \sum_{j=1}^{n} \left|\sin\left(\frac{j\pi}{2}\right) - \sin\left(\frac{(j+1)\pi}{2}\right)\right|.$$

Now, if j is even, then

$$\left|\sin\left(\frac{j\pi}{2}\right) - \sin\left(\frac{(j+1)\pi}{2}\right)\right| = |0 - (\pm 1)| = 1,$$

while if j is odd,

$$\left|\sin\left(\frac{j\pi}{2}\right) - \sin\left(\frac{(j+1)\pi}{2}\right)\right| = |(\pm 1) - 0| = 1.$$

Therefore,

$$V_S(f, (0, 1)) = \sum_{j=1}^{n} 1 = n,$$

and so for S of this form we have $V_S(f, (0, 1)) \to \infty$ as $n \to \infty$. It follows that $V(f, (0, 1)) = \infty$.

If $V(f, I)$ is *finite* for a particular function $f : I \to \mathbb{R}$, we say that f has **bounded variation** (or is a **function of bounded variation**) on I. Example 2-7-1 shows that all step functions on I have bounded variation on I, while Example 2-7-2 is an example of a function that does not have bounded variation.

There is a very important connection between functions of bounded variation and monotone functions, which we must now discuss. Recall first that $f : I \to \mathbb{R}$ is *monotone increasing* on I if $f(x_1) \leq f(x_2)$ whenever $x_1 < x_2$ $(x_1, x_2 \in I)$, and *monotone decreasing* on I if $f(x_1) \geq f(x_2)$ whenever $x_1 < x_2$ $(x_1, x_2 \in I)$; in either case we say that f is *monotone* on I. A very slight modification of the appropriate part of the proof of Theorem 2.4.1 shows that if f is monotone on I, where I is an interval with endpoints a, b, then $f(t^-)$ and $f(t^+)$ exist and are finite for all t such that $a < t < b$, and also $f(a^+)$ and $f(b^-)$ exist (but are not necessarily finite). Furthermore, $f(a^+)$ and $f(b^-)$ are both finite if and only if $\sup\{f(x) : x \in I\}$ and $\inf\{f(x) : x \in I\}$ are both finite.

Lemma 2.7.1
Let I be an interval, and $f : I \to \mathbb{R}$ a function of bounded variation on I. For any $x \in I$, denote by I_x the interval $\{t : t \in I, t \leq x\} \subseteq I$. Then
- *(i) $0 \leq V(f, I_x) \leq V(f, I)$ for all $x \in I$;*
- *(ii) the function $g : I \to \mathbb{R}$ defined by $g(x) = V(f, I_x)$ for all $x \in I$ is monotone increasing on I.*

Proof Part (i) follows at once from the result proved in Exercises 1-3, No. 3, and the fact that any partial subdivision of I_x is also a partial subdivision of I. To prove part (ii), let $x_1, x_2 \in I$ be such that $x_1 < x_2$. Then $I_{x_1} \subseteq I_{x_2}$, so any partial subdivision of I_{x_1} is also a partial subdivision of I_{x_2}. Thus $V(f, I_{x_1}) \leq V(f, I_{x_2})$, which proves that g is monotone increasing on I. □

Theorem 2.7.2
Let I be any interval. Then a function $f : I \to \mathbb{R}$ has bounded variation on I if and only if f can be expressed as a difference

$$f = h_1 - h_2,$$

where the functions $h_1, h_2 : I \to \mathbb{R}$ are both monotone increasing on I, and $\sup\{h_1(x) : x \in I\}$, $\inf\{h_1(x) : x \in I\}$, $\sup\{h_2(x) : x \in I\}$, $\inf\{h_2(x) : x \in I\}$ are all finite.

Proof We prove first that if f has bounded variation on I, then it can be represented by the difference $h_1 - h_2$, where the h_k are as claimed in the theorem. Suppose f has bounded variation on I. For each $x \in I$, define I_x as in Lemma 2.7.1. Define $h_1, h_2 : I \to \mathbb{R}$ by $h_1(x) = V(f, I_x)$ and $h_2(x) = V(f, I_x) - f(x)$ for each $x \in I$. Then certainly $f = h_1 - h_2$, and Lemma 2.7.1 shows that h_1 has all the required properties.

Also, if $x_1, x_2 \in I$ are such that $x_1 < x_2$, then

$$h_2(x_2) - h_2(x_1) = V(f, I_{x_2}) - V(f, I_{x_1}) - [f(x_2) - f(x_1)]. \tag{2.2}$$

Now, if S is any partial subdivision of I_{x_1}, then $S^* = S \cup \{[x_1, x_2]\}$ is a partial subdivision of I_{x_2}, and

$$V_S(f, I_{x_1}) + |f(x_2) - f(x_1)| = V_{S^*}(f, I_{x_2}) \le V(f, I_{x_2}).$$

Thus, $V_S(f, I_{x_1}) \le V(f, I_{x_2}) - |f(x_2) - f(x_1)|$ for all partial subdivisions S of I_{x_1}, and so

$$\begin{aligned} V(f, I_{x_1}) &= \sup\{V_S(f, I_{x_1}) : S \text{ a partial subdivision of } I_{x_1}\} \\ &\le V(f, I_{x_2}) - |f(x_2) - f(x_1)|. \end{aligned}$$

Hence $V(f, I_{x_2}) - V(f, I_{x_1}) \ge |f(x_2) - f(x_1)| \ge f(x_2) - f(x_1)$, and it follows from equation (2.2) that $h_2(x_2) - h_2(x_1) \ge 0$, so h_2 is monotone increasing on I.

Finally, it can be shown (see Exercises 2-7, No. 1) that $M = \sup\{f(x) : x \in I\}$ and $m = \inf\{f(x) : x \in I\}$ are finite, and so by virtue of Lemma 2.7.1(i) we have that for any $x \in I$, $m \le f(x) \le M$ and $0 \le V(f, I_x) \le V(f, I)$, i.e., $-M \le -f(x) \le -m$ and $0 \le V(f, I_x) \le V(f, I)$, and therefore $-M \le V(f, I_x) - f(x) \le V(f, I) - m$, i.e., $-M \le h_2(x) \le V(f, I) - m$, where $-M$ and $V(f, I) - m$ are both finite. It follows that $\sup\{h_2(x) : x \in I\}$ and $\inf\{h_2(x) : x \in I\}$ are both finite as required.

It remains to show that if $f = h_1 - h_2$, where h_1 and h_2 satisfy the conditions prescribed in the theorem, then f has bounded variation on I. We leave the proof of this as an exercise. □

Corollary 2.7.3

If I is an interval with endpoints a, b and $f : I \to \mathbb{R}$ has bounded variation on I, then $f(t^-)$ and $f(t^+)$ exist and are finite for all t such that $a < t < b$, and also $f(a^+)$ and $f(b^-)$ exist and are finite.

Proof This follows at once from Theorem 2.7.2, the corresponding properties of monotone functions, and the basic rules for limits. □

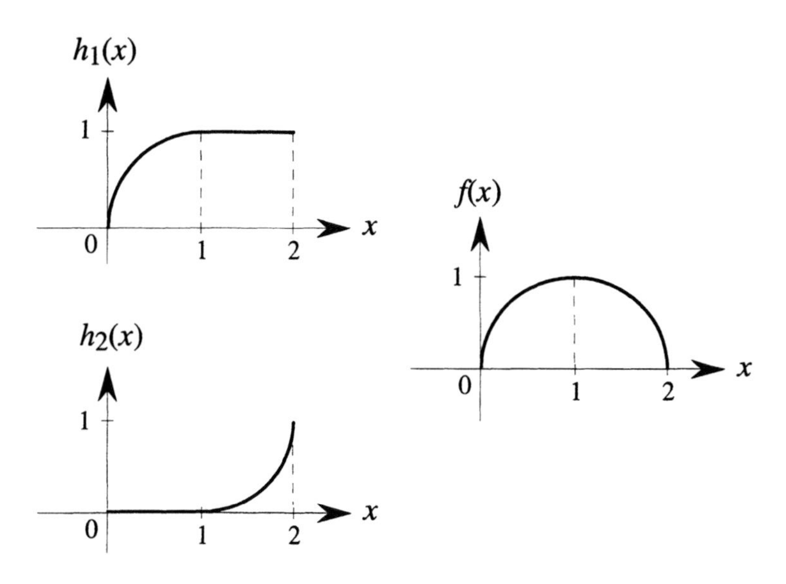

FIGURE 2.14

Example 2-7-3:
Let $f(x) = \sqrt{2x - x^2}$ on the interval $[0, 2]$. Define the functions $h_1, h_2 : [0, 2] \to \mathbb{R}$ by

$$h_1(x) = \begin{cases} \sqrt{2x - x^2}, & \text{if } 0 \leq x \leq 1, \\ 1, & \text{if } 1 < x \leq 2, \end{cases}$$

$$h_2(x) = \begin{cases} 0, & \text{if } 0 \leq x \leq 1, \\ 1 - \sqrt{2x - x^2} & \text{if } 1 < x \leq 2. \end{cases}$$

Then $f = h_1 - h_2$ on I, and the conditions prescribed in Theorem 2.7.2 are certainly satisfied by h_1 and h_2, so f has bounded variation on $[0, 2]$ (cf. Figure 2.14).

Note that the expression of a particular function of bounded variation as a difference of monotone increasing functions is by no means unique. For instance, just replacing h_1 and h_2 by $h_1 + k$ and $h_2 + k$, where k is a constant, gives an infinite number of different expressions of this kind; there are other possibilities as well.

There is no necessary connection between continuity of a function and the property of having bounded variation. Since all step functions have bounded variation (see Example 2-7-1), functions of bounded variation need not be continuous. Conversely, a continuous function need not have bounded variation. For instance, the function $\tan(x)$ is monotone increasing and continuous on $(-\pi/2, \pi/2)$, but since its set of values does not have finite upper and lower bounds, it does not have bounded variation on this interval (See Exercises 2-7, No. 3). Indeed, Example 2-7-2 shows that even if the set of values of a continuous function has finite upper and lower bounds, the function need not have bounded variation (see also Exercises 2-7, No. 5).

We say that a function $f :\to \mathbb{R}$ is **absolutely continuous** on I if given any $\epsilon > 0$, there exists a $\delta > 0$ (depending on ϵ) such that $V_S(f, I) < \epsilon$ for all partial subdivisions S of I for which the sum of the lengths of all the constituent intervals is less than δ. It follows easily, by considering only partial subdivisions consisting of a single interval, that if a function is absolutely continuous on I, it is also continuous on I. We have also the following theorem:

Theorem 2.7.4
If I is an interval with **finite** *endpoints a, b and $f : I \to \mathbb{R}$ is absolutely continuous on I, then f has bounded variation on I.*

Proof Let $S = \{I_1, I_2, \ldots, I_n\}$ be any partial subdivision of I such that $I_j = [a_j, b_j]$ for $j = 1, 2, \ldots, n$. Let $\epsilon = 1$ in the definition of absolute continuity. Then there exists a $\delta_1 > 0$ such that $V_{S^*}(f, I) < 1$ for all partial subdivisions S^* of I for which the sum of the lengths of all the constituent intervals is less than δ_1. Let N be the smallest positive integer greater than $(b - a)/\delta_1$. Then $1/N < \delta_1/(b - a)$.

Now take any $j = 1, 2, \ldots, n$. Divide I_j into N subintervals of equal length,

$$I_{j1} = [a_j(= x_{j0}), x_{j1}],\ I_{j2} = [x_{j1}, x_{j2}],\ \ldots,\ I_{jN} = [x_{j(N-1)}, b_j(= x_{jN})],$$

and denote the length of I_{jr} by ℓ_{jr} ($j = 1, 2, \ldots, n; r = 1, 2, \ldots, N$). Then

$$\ell_{jr} = \frac{b_j - a_j}{N} < \frac{\delta_1(b_j - a_j)}{b - a}.$$

For each $r = 1, 2, \ldots, N$, let $S_r = \{I_{1r}, I_{2r}, \ldots, I_{nr}\}$. Then S_r is a partial subdivision of I, and the sum of the lengths of all its constituent intervals is

$$\sum_{j=1}^{n} \ell_{jr} < \frac{\delta_1}{b-a} \sum_{j=1}^{n} (b_j - a_j) \le \delta_1,$$

since $\sum_{j=1}^{n}(b_j - a_j)$ is the sum of the lengths of all the intervals that make up the partial subdivision S of I, and is therefore no greater than the length $b - a$ of I. Therefore, $V_{S_r}(f, I) < 1$ for each $r = 1, 2, \ldots, N$.

Now

$$\begin{aligned} V_S(f, I) &= \sum_{j=1}^{n} |f(b_j) - f(a_j)| \\ &= \sum_{j=1}^{n} |f(x_{jN}) - f(x_{j(N-1)}) + f(x_{j(N-1)}) - \cdots + f(x_{j2}) - f(x_{j1}) \\ &\qquad + f(x_{j1}) - f(x_{j0})| \\ &\le \sum_{j=1}^{n} \sum_{r=1}^{N} |f(x_{jr}) - f(x_{j(r-1)})| \quad \text{(by the triangle inequality)} \\ &= \sum_{r=1}^{N} \sum_{j=1}^{n} |f(x_{jr}) - f(x_{j(r-1)})| \\ &= \sum_{r=1}^{N} V_{S_r}(f, I) \le \sum_{r=1}^{N} 1 = N. \end{aligned}$$

Since N is finite and independent of S, it follows that $V(f, I) \le N$ and f has bounded variation on I. □

Note that the converse of this theorem does not hold; as we have seen earlier, a function of bounded variation need not even be continuous, let alone absolutely continuous.

Note also that we have seen examples of continuous functions on finite intervals that do not have bounded variation, and therefore (by Theorem 2.7.4) are not absolutely continuous. Thus absolute continuity, as the name suggests, is a stronger condition than continuity, in the sense that the set of absolutely continuous functions on an interval is a proper subset of the set of continuous functions on that interval. Some examples of functions that *are* absolutely continuous are given in Exercises 2-7, No. 6.

Exercises 2-7:

1. Prove that if $f : I \to \mathbb{R}$ has bounded variation on I, then $\sup\{f(x) : x \in I\}$ and $\inf\{f(x) : x \in I\}$ are both finite.
2. Prove that if $f, g : I \to \mathbb{R}$ have bounded variation on I, then so do kf (where k is a real number), $f + g$, and fg.
3. Prove that a function $f : I \to \mathbb{R}$ that is monotone on I has bounded variation on I if and only if $\sup\{f(x) : x \in I\}$ and $\inf\{f(x) : x \in I\}$ are both finite. Use this, together with the results proved in the preceding exercise, to prove part (b) of Theorem 2.7.2.
4. Express the step function θ defined in Example 2-7-1 as a difference of two monotone increasing functions, and sketch the graphs of the two functions.
5. Let $f : [-\pi/2, \pi/2] \to \mathbb{R}$ be defined by

$$f(x) = \begin{cases} x\sin(\frac{1}{x}), & \text{if } x \neq 0, \\ 0, & \text{if } x = 0. \end{cases}$$

   (a) Prove that f is continuous on $[-\pi/2, \pi/2]$; you may assume that $x\sin(1/x)$ is continuous for all $x \neq 0$, so all that has to be proved is that f is continuous at $x = 0$.

   (b) Use a method similar to that used in Example 2-7-2 to show that f does not have bounded variation on $[-\pi/2, \pi/2]$.
6. A function $f : I \to \mathbb{R}$ is said to be a *Lipschitz function* on I if there exists a real number L such that $|f(x_1) - f(x_2)| \leq L|x_1 - x_2|$ for all $x_1, x_2 \in I$.

   (a) Prove that any Lipschitz function on I is absolutely continuous on I.

   (b) Use this result to show that any linear function is absolutely continuous on any interval and that the function x^2 is absolutely continuous on any interval with finite endpoints.

   (c) Use a proofby contradiction to show that x^2 is not absolutely continuous on $(-\infty, \infty)$.

3 CHAPTER The Riemann Integral

The development of a rigorous theory of the definite integral in the nineteenth century is associated particularly with the work of Augustin-Louis Cauchy (1789–1857) in France, and Bernhard Riemann (1826–1866) in Germany. In order to give some background to the modern theory, we will describe briefly a definition of an integral equivalent to that introduced by Riemann in 1854, and discuss some of the weaknesses in this definition that suggest the need for a more general theory.

3.1 Definition of the Integral

Let $[a, b]$ be any closed interval. A function $f : [a, b] \to \mathbb{R}$ is said to be **Riemann integrable** over $[a, b]$ if and only if for any number $\epsilon > 0$, there exist step functions $g^\epsilon, G^\epsilon : [a, b] \to \mathbb{R}$ such that

(i) $g^\epsilon \leq f \leq G^\epsilon$;

(ii) $A(G^\epsilon) - A(g^\epsilon) \leq \epsilon$

(cf. Figure 3.1). A function $f : I \to \mathbb{R}$ for which the set $\{f(x) : x \in I\}$ has finite upper and lower bounds will be called **bounded** on I.

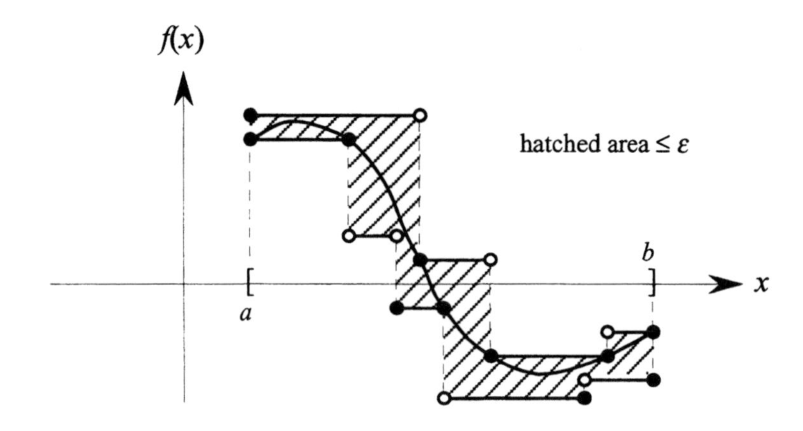

FIGURE 3.1

From condition (i) it is clear that any Riemann integrable function over $[a, b]$ must be bounded on $[a, b]$, and that

$$A(f) = \sup\{A(g) : g : [a, b] \to \mathbb{R} \text{ is a step function and } g \leq f \text{ on } [a, b]\}$$

is finite. We call $A(f)$ the **Riemann integral** of f over $[a, b]$, and it is usually denoted by $\int_a^b f(x)\,dx$. Note that if a function $f : [a, b] \to \mathbb{R}$ is Riemann integrable, then condition (ii) implies that

$$A(f) = \inf\{A(G) : G : [a, b] \to \mathbb{R} \text{ is a step function and } G \geq f \text{ on } [a, b]\}.$$

Riemann's definition extends the concept of the "area under the graph" of f to a wider class of functions than step functions by using step functions to approximate f. This is essentially the definition of the integral that is used in elementary calculus, and its properties are familiar to all students of the calculus.

The main focus of our study will be the Lebesgue–Stieltjes integral, which is a generalization of the Riemann integral. The Riemann integral, however, is still important, because many calculations involving the Lebesgue–Stieltjes integral involve the Riemann integral. Moreover, one must understand some basic facts about the Riemann integral in order to understand the relationship between the two integrals and appreciate the need for a more general theory. We will not embark on a detailed account of the Riemann integral. Instead, we will limit ourselves to a brief discussion of the integral, highlighting any properties that are of use later in establishing the relationship between the Riemann and Lebesgue–Stieltjes integrals.

Given a bounded function $f : I \to \mathbb{R}$, candidates for the step functions g^ϵ, G^ϵ in the above definition can be constucted by partitioning the interval and defining step functions based on the maximum and minimum values the functions assume in the subintervals. By a **partition** P of an interval $I = [a, b]$ we mean a *finite* set of numbers $x_0, x_1, \cdots, x_n$, where

$$a = x_0 < x_1 < \cdots < x_n = b.$$

Let $I_1 = [x_0, x_1]$, and for $1 < k \leq n$ let $I_k = (x_{k-1}, x_k]$ denote the kth subinterval of I associated with the partition P. Let $\Delta_k = x_k - x_{k-1}$ denote the length of the subinterval. If $f : I \to \mathbb{R}$ is bounded on I, then given any partition P of I, step functions $g_P, G_P : I \to \mathbb{R}$ such that $g_P \leq f \leq G_P$ on I can readily be constructed. Let

$$M_k = \sup_{x \in I_k} f(x), \quad m_k = \inf_{x \in I_k} f(x),$$

and define g_P, G_P as follows:

$$g_P(x) = \begin{cases} m_1, & \text{if } x \in I_1, \\ m_2, & \text{if } x \in I_2, \\ \vdots & \vdots \\ m_n, & \text{if } x \in I_n, \end{cases} \qquad G_P(x) = \begin{cases} M_1, & \text{if } x \in I_1, \\ M_2, & \text{if } x \in I_2, \\ \vdots & \vdots \\ M_n, & \text{if } x \in I_n \end{cases} \tag{3.1}$$

(cf. Figure 3.2). Let $\overline{S}_P(f) = A(G_P) = \sum_{k=1}^{n} M_k \Delta_k$ and $\underline{S}_P(f) = A(g_P) = \sum_{k=1}^{n} m_k \Delta_k$; evidently, $\overline{S}_P(f) \geq \underline{S}_P(f)$ for any partition P of I. A partition P' of I is called a **refinement** of the partition P of I if every x_k in P corresponds to some x'_j in P'. Thus a refinement P' of P can be constructed from P by distributing additional partition points between those already occurring in P.

Lemma 3.1.1

If $f : [a, b] \to \mathbb{R}$ is bounded on $I = [a, b]$ and if P is a partition of I, then for any refinement P' of P,

$$\overline{S}_{P'}(f) \leq \overline{S}_P(f)$$

and

$$\underline{S}_{P'}(f) \geq \underline{S}_P(f).$$

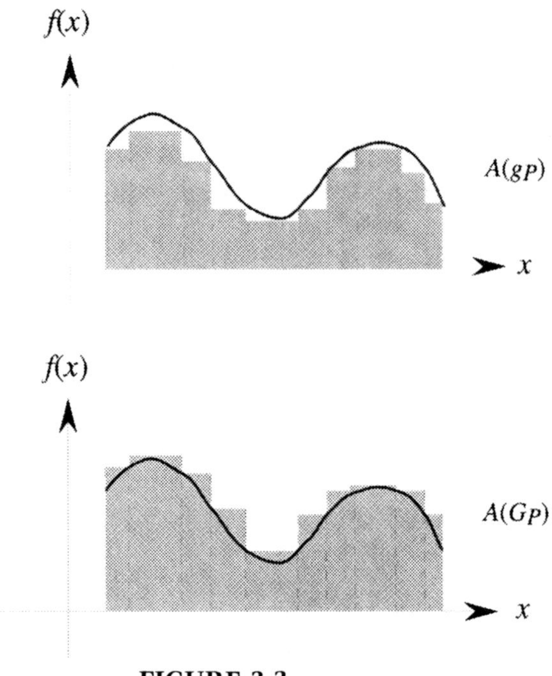

FIGURE 3.2

Lemma 3.1.2

Suppose that $f : [a, b] \to \mathbb{R}$ is bounded on $I = [a, b]$ and that P and P' are any two partitions of I. Then

$$\overline{S}_P(f) \geq \underline{S}_{P'}(f).$$

Let Π denote the set of all partitions of $I = [a, b]$. From Lemmas 3.1.1 and 3.1.2 the set $\overline{S} = \{\overline{S}_P(f) : P \in \Pi\}$ is bounded below by $\underline{S}_I(f)$, where I is the partition corresponding to $x_0 = a,\ x_1 = b$. The set $\overline{S}$ must therefore have a finite lower bound if f is bounded on I. Similarly, if f is bounded on I, then the set $\underline{S} = \{\underline{S}_P(f) : P \in \Pi\}$ must have a finite upper bound, because it is bounded above by $\overline{S}_I(f)$. The quantities $\inf_{P\in\Pi} \overline{S}_P(f)$ and $\sup_{P\in\Pi} \underline{S}_P(f)$ are finite and are called the upper and lower Riemann–Darboux integrals of f over I, respectively. If, in addition, it is assumed that f is Riemann integrable over

I, then it can be shown that

$$A(f) = \inf_{P\in\Pi} \overline{S}_P(f) = \sup_{P\in\Pi} \underline{S}_P(f).$$

Indeed, the condition $\inf_{P\in\Pi} \overline{S}_P(f) = \sup_{P\in\Pi} \underline{S}_P(f)$ is commonly used in the definition of a Riemann integrable function.

If $f : [a, b] \to \mathbb{R}$ is Riemann integrable on $I = [a, b]$, then there is a sequence of partitions $\{\overline{P}_j\}$, $\overline{P}_j \in \Pi$, such that $\lim_{j\to\infty} \overline{S}_{\overline{P}_j}(f) = A(f)$ and a sequence $\{\underline{P}_k\}$, $\underline{P}_k \in \Pi$, such that $\lim_{k\to\infty} \underline{S}_{\underline{P}_k}(f) = A(f)$. Now, for any two partitions P and P', the set $P \cup P'$ yields a partition Q that is the common refinement of P and P'. Lemma 3.1.1 and the definitions of infimum and supremum thus indicate that we can always find a sequence $\{P_k\}$ such that $\lim_{k\to\infty} \overline{S}_{P_k}(f) = \lim_{k\to\infty} \underline{S}_{P_k}(f) = A(f)$, and moreover, we can assume that P_{k+1} is a refinement of $P_k, k = 1, 2, \ldots$.

Theorem 3.1.3

If $f : [a, b] \to \mathbb{R}$ is Riemann integrable over $I = [a, b]$, then there exists a sequence of partitions $\{P_k\}$, $P_k \in \Pi$, such that P_{k+1} is a refinement of P_k, $k = 1, 2, \ldots$, and

$$\lim_{k\to\infty} \overline{S}_{P_k}(f) = \lim_{k\to\infty} \underline{S}_{P_k}(f) = A(f).$$

Let $P = \{x_0, x_1, \ldots, x_n\}$ be a partition of the interval $I = [a, b]$. The **norm** of P, denoted by $\|P\|$, is defined as

$$\|P\| = \max_{k=1,2,\ldots,n} \Delta_k.$$

The norm of P is thus the maximum of all the lengths of all the subintervals formed by the partition P. It can be shown that if $f : I \to \mathbb{R}$ is Riemann integrable over I, then any sequence of partitions $\{\hat{P}_j\}$ such that $\|\hat{P}_j\| \to 0$ as $j \to \infty$ will produce the Riemann integral of f over I, i.e.,

$$\lim_{j\to\infty} \overline{S}_{\hat{P}_j}(f) = \lim_{j\to\infty} \underline{S}_{\hat{P}_j}(f) = \int_a^b f\, dx.$$

In general, it is not particularly convenient to prove that a given function is Riemann integrable directly from the definition. The following theorems are thus useful in this regard:

Theorem 3.1.4
If $f : [a, b] \to \mathbb{R}$ is monotone on $I = [a, b]$, then it is Riemann integrable over I.

Theorem 3.1.5
If $f : [a, b] \to \mathbb{R}$ is continuous on $I = [a, b]$, then it is Riemann integrable over I.

Exercises 3-1:

1. Let $f : [a, b] \to \mathbb{R}$ be a bounded function on $I = [a, b]$ and let P' be a refinement of the partition P of I. Prove that $g_P \leq g_{P'}$ and $G_P \geq G_{P'}$, where the step functions g and G are as defined in equation (3.1).
2. Use Lemma 3.1.1 and the fact that $Q = P \cup P'$ is a common refinement for any two partitions P and P' of I to prove Lemma 3.1.2.

3.2 Improper Integrals

The Riemann integral as defined in Section 3-1 is over *closed* intervals. The defintion of the integral can be extended to other intervals by using a limiting process leading to the theory of what are usually called "improper integrals." We eschew a detailed account of improper integrals; instead, we give a brief description of the basic idea with examples.

Suppose that f is a continuous function on the interval $(a, b]$. By Theorem 3.1.5 the function f is Riemann integrable over any interval of the form $[c, b]$, where $a < c < b$, and we can enquire about the existence of $\lim_{c \to a^+} \int_c^b f(x)\,dx$. If this limit is finite, then we say that it defines the improper integral of f from a to b. The improper integral is denoted in the same way as the Riemann integral, i.e., by $\int_a^b f(x)\,dx$. If an improper integral exists, we also say that it **converges**. Improper integrals over other intervals such as $[a, b)$, $[a, \infty)$, $(-\infty, b]$, etc. are defined in a similar way.

Example 3-2-1:
The function f defined by $f(x) = 1/\sqrt{x}$ is continuous for all $x \in (0, 1]$. By Theorem 3.1.5 f is Riemann integrable in any closed subset of

$(0, 1]$. In fact, for any $0 < c < 1$,

$$\int_c^1 f(x)\,dx = \left[2\sqrt{x}\right]_c^1 = 2(1 - \sqrt{c}).$$

Now, $\lim_{c\to 0^+} \int_c^1 f(x)\,dx = 2(1 - \lim_{c\to 0^+} \sqrt{c}) = 2$, and therefore the improper integral $\int_0^1 f(x)\,dx$ exists.

Example 3-2-2:
The function f defined by $f(x) = 1/x^2$ does not have an improper integral from 0 to 1. The function f is Riemann integrable over any interval $[c, 1]$, $0 < c < 1$, because it is continuous there, but

$$\int_c^1 \frac{1}{x^2}\,dx = \left[-\frac{1}{x}\right]_c^1 = \frac{1}{c} - 1,$$

and thus $\lim_{c\to 0^+} \int_c^1 f(x)\,dx$ is not finite.

Example 3-2-3:
Let f be the function defined in Example 3-2-2 and consider the interval $[1, \infty)$. The function f is Riemann integrable over any interval of the form $[1, c]$, where $1 < c < \infty$, and since

$$\lim_{c\to\infty} \int_1^c \frac{1}{x^2}\,dx = \lim_{c\to\infty} \left[-\frac{1}{x}\right]_1^c = \lim_{c\to\infty} \left(1 - \frac{1}{c}\right) = 1,$$

the improper integral $\int_1^\infty 1/x^2 dx$ converges.

Example 3-2-4:
Let $f : [1, \infty) \to \mathbb{R}$ be defined by $f(x) = 1/x$. In any closed interval $[1, c]$, $c > 1$, we have that

$$\int_1^c \frac{1}{x}\,dx = \left[\log x\right]_1^c = \log c.$$

Since $\log c \to \infty$ as $c \to \infty$, the improper integral $\int_1^\infty 1/x\,dx$ does not exist, i.e., it diverges.

Although the definition of the Riemann integral can be extended to open or semiopen intervals, many of the results concerning the Riemann integral over a closed interval do not carry over in the extension. Example 3-2-2 indicates that continuity on a semiopen interval does not guarantee the existence of the improper integral.

Examples 3-2-2 and 3-2-4 show that monotonicity does not imply Riemann integrability when the interval is not closed.

If $f : [a, b] \to \mathbb{R}$ is Riemann integrable over $[a, b]$, then it can be shown that $|f|$ is also Riemann integrable over $[a, b]$. For improper integrals, this is no longer true, i.e., $\int_a^b f(x)\,dx$ may converge but $\int_a^b |f(x)|\,dx$ may diverge. If $\int_a^b f(x)\,dx$ and $\int_a^b |f(x)|\,dx$ both converge, then the improper integral is called **absolutely convergent**. If $\int_a^b f(x)\,dx$ converges but $\int_a^b |f(x)|\,dx$ diverges, then the improper integral is called **conditionally convergent**. The integral in Example 3-2-3 is absolutely convergent. The next example requires more familiarity with improper integrals than assumed heretofore, but it provides a specific example of a conditionally convergent integral.

Example 3-2-5:
Let $f : [\pi, \infty) \to \mathbb{R}$ be defined by $f(x) = (\sin x)/x$. Over any closed interval of the form $[\pi, c]$, $c > \pi$, the function f is Riemann integrable, and (anticipating integration by parts) we have

$$\int_\pi^c \frac{\sin x}{x}\,dx = \left[-\frac{\cos x}{x}\right]_\pi^c + \int_\pi^c \frac{\cos x}{x^2}\,dx.$$

Now, $|\cos x/x^2| \leq 1/x^2$ for all $x \in [\pi, c]$, and it can be shown that $\lim_{c\to\infty} \int_\pi^c \cos x/x^2\,dx$ exists, since $\lim_{c\to\infty} \int_\pi^c 1/x^2\,dx$ exists (the comparison test). On the other hand, it can be shown that $\lim_{c\to\infty} \int_\pi^c |\sin x/x|dx$ does not exist.

The definition of the Riemann integral can thus be extended to intervals of integration other than closed intervals by using improper integrals. The modern approach, to be described in the next chapter, works with arbitrary intervals from the start, leading to a tidier theory, but this is a relatively minor improvement. A more fundamental weakness of Riemann's approach is revealed in the next section.

3.3 A Nonintegrable Function

Theorems 3.1.4 and 3.1.5 indicate that the class of Riemann integrable functions is a large one. In fact, it can be proved that if $f : [a, b] \to \mathbb{R}$ is bounded on $[a, b]$ and the set of all points of discontinuity of f in $[a, b]$ is either empty, finite, or countably infinite, then f is Riemann integrable over $[a, b]$. However, if the set of points of discontinuity of f is infinite but not countable, then f may not be Riemann integrable, as the following example illustrates.

Example 3-3-1:
Let $f : [0, 1] \to \mathbb{R}$ be defined by

$$f(x) = \begin{cases} 1, & \text{if } x \text{ is rational, } x \neq 0, 1, \\ 0, & \text{if } x \text{ is irrational or } x = 0 \text{ or } x = 1. \end{cases}$$

Suppose that f were Riemann integrable over $[0, 1]$. Then taking $\epsilon = \frac{1}{2}$ in the definition of Riemann integrability, there must exist step functions $g, G : [0, 1] \to \mathbb{R}$ such that $g \leq f \leq G$ on $[0, 1]$ and $A(G) - A(g) \leq \frac{1}{2}$.

Now, we have seen that any interval of nonzero length contains infinitely many rational numbers and infinitely many irrational numbers. Thus we must have $g(x) \leq 0$ and $G(x) \geq 1$ for all but a finite number (possibly zero) of points $x \in [0, 1]$. The values of $g(x)$ and $G(x)$ at a finite number of values of x do not affect the values of the areas $A(g)$ and $A(G)$, so we must have $A(g) \leq 0$ and $A(G) \geq 1$. Thus $A(G) - A(g) \geq 1$, which contradicts $A(G) - A(g) \leq \frac{1}{2}$, and so f cannot be Riemann integrable over $[0, 1]$. Note that f is discontinuous at *every* point in the interval $[0, 1]$.

The reader might rightly ask why we should be concerned that this rather peculiar function does not have an integral in the Riemann sense. The reason is connected with the following concern: Suppose the functions $f_n : [a, b] \to \mathbb{R}$ are Riemann integrable over $[a, b]$ for all $n = 1, 2, \ldots$, and $f_n \to f$ on $[a, b]$. It is natural to hope that the property of integrability "carries over to the limit," so that we can be sure that f is also Riemann integrable over $[a, b]$ (and that the integral of f_n tends to the integral of f as n tends to ∞), but this may not be the case. This concern is important, because the solutions to many problems in the calculus such as differential equations are often obtained as the limit of a sequence of successive approximations. Unfortunately, there are sequences of functions that are Riemann integrable but that converge to a function that is not. It is necessary to find only one counterexample to destroy our hopes.

We will now show that the nonintegrable function defined in the previous example is the limit of a sequence of Riemann integrable functions. We know that the rational numbers in $(0, 1)$ form a countably infinite set, so we can write the set of rationals in $(0, 1)$ as $\{r_1, r_2, r_3, \ldots\}$. For each $j = 1, 2, \ldots$ we subdivide $[0, 1]$ into three subintervals: $I_{j1} = [0, r_j)$, $I_{j2} = [r_j, r_j]$, and $I_{j3} = (r_j, 1]$. We then define $\theta_j : [0, 1] \to \mathbb{R}$ by

$$\theta_j(x) = \begin{cases} 0, & \text{if } x \in I_{j1}, \\ 1, & \text{if } x \in I_{j2}, \\ 0, & \text{if } x \in I_{j3}. \end{cases}$$

Now, θ_j is a step function for each $j = 1, 2, \ldots$. Define $f_n : [0, 1] \to \mathbb{R}$ by

$$f_n = \sum_{j=1}^{n} \theta_j.$$

Now, each f_n is also a step function, and is therefore Riemann integrable over $[0, 1]$. Furthermore, it is evident that

$$f_n(x) = \begin{cases} 0, & \text{if } x \text{ is irrational or } x \in \{0, 1, r_{n+1}, r_{n+2} \ldots\}, \\ 1, & \text{if } x \in \{r_1, r_2, \ldots, r_n\}. \end{cases}$$

Thus if $x \in [0, 1]$ is irrational or $x = 0$ or $x = 1$, then $f_n(x) = 0 = f(x)$ for all $n = 1, 2, \ldots$, and so $f_n(x) \to f(x)$ as $n \to \infty$. If $x \in [0, 1]$ is rational, say $x = r_N$ $(N = 1, 2, \ldots)$, then $f_n(x) = 1 = f(x)$ for all $n \geq N$, so again $f_n(x) \to f(x)$ as $n \to \infty$. We have therefore established that the Riemann integrable sequence of functions f_n converges to the nonintegrable function f on $[0, 1]$.

This example shows that integrability in the Riemann sense does not always carry over in the limit. It can be shown that it does under certain conditions, but these conditions are rather complicated. Because the modern theory (as we shall see) allows for a wider class of integrable functions, the conditions under which integrability carries over to the limit are much simpler and easier to use. This is one of the most important ways in which the modern theory is an improvement over the older one.

4 CHAPTER The Lebesgue-Stieltjes Integral

We now proceed to formulate the definition of the integral that we are going to study. It results from combining the ideas of two people. The French mathematician Henri Lebesgue (1875–1941), building on earlier work by Emile Borel (1871–1956) on the measure of a set, succeeded in defining an integral (the Lebesgue integral) that applied to a wider class of functions than did the Riemann integral, and for which the convergence theorems were much simpler. The Dutch mathematician Thomas Stieltjes (1856–1894) was responsible for the notion of integrating one function with respect to another function. His ideas were originally developed as an extension of the Riemann integral, known as the Riemann–Stieltjes integral. The subsequent combination of his ideas with the measure-theoretic approach of Lebesgue has resulted in a very powerful and flexible concept of integration.

4.1 The Measure of an Interval

Let $\alpha : \mathbb{R} \to \mathbb{R}$ be a monotone increasing function, and let I be an interval with endpoints a, b. We define the **α-measure** of I, denoted by $\mu_\alpha(I)$, as follows:

$$\mu_\alpha([a,b]) = \alpha(b^+) - \alpha(a^-),$$
$$\mu_\alpha((a,b]) = \alpha(b^+) - \alpha(a^+),$$
$$\mu_\alpha([a,b)) = \alpha(b^-) - \alpha(a^-),$$

and if $a < b$,

$$\mu_\alpha((a,b)) = \alpha(b^-) - \alpha(a^+).$$

The "open interval" (a,a) is of course the empty set, and we define $\mu_\alpha((a,a))$ to be zero for any $a \in \mathbb{R}_e$. The intervals $(a,a]$ and $[a,a)$ are also empty, but in those cases the fact that their α-measure is zero follows from the general definition, and need not be specified separately.

It follows easily from Theorem 2.4.1(i) and Corollary 2.4.2(i) that $\mu_\alpha(I) \geq 0$ for any interval I, and that if I and J are intervals with $I \subseteq J$, then $\mu_\alpha(I) \leq \mu_\alpha(J)$.

If a and b are finite, and α is continuous at both a and b, then we have $\alpha(a^-) = \alpha(a^+) = \alpha(a)$ and $\alpha(b^-) = \alpha(b^+) = \alpha(b)$, and so $\mu_\alpha(I) = \alpha(b) - \alpha(a)$ in all four cases. In particular, if $\alpha(x) = x$ for all $x \in \mathbb{R}$, then $\mu_\alpha(I) = b - a$ is the ordinary length of the interval I. In general, the α-measure of an interval is just the change in the value of α over the interval in question; it can be thought of as a generalization of the notion of length.

Example 4-1-1:
Let $\alpha : \mathbb{R} \to \mathbb{R}$ be defined by

$$\alpha(x) = \begin{cases} 0, & \text{if } x < 1, \\ x^2 - 2x + 2, & \text{if } 1 \leq x < 2, \\ 3, & \text{if } x = 2, \\ x + 2, & \text{if } x > 2 \end{cases}$$

(cf. Figure 4.1). Then:

$$\mu_\alpha([1,2]) = \alpha(2^+) - \alpha(1^-) = 4 - 0 = 4,$$
$$\mu_\alpha((1,2]) = \alpha(2^+) - \alpha(1^+) = 4 - 1 = 3,$$
$$\mu_\alpha([1,2)) = \alpha(2^-) - \alpha(1^-) = 2 - 0 = 2,$$

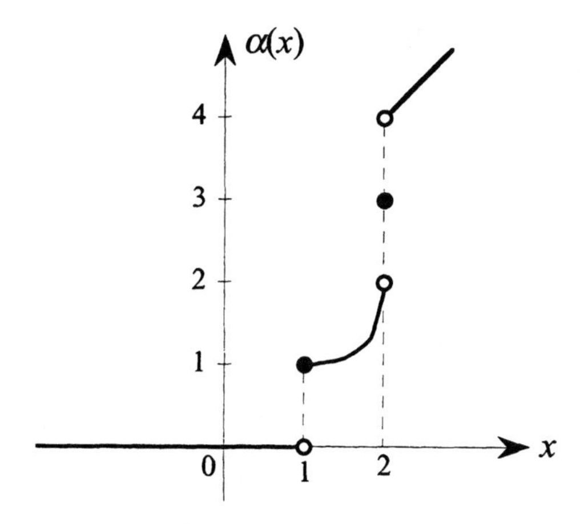

FIGURE 4.1

$$\begin{aligned}
\mu_\alpha((1,2)) &= \alpha(2^-) - \alpha(1^+) = 2 - 1 = 1,\\
\mu_\alpha([2,3]) &= \alpha(3^+) - \alpha(2^-) = 5 - 2 = 3,\\
\mu_\alpha((2,3)) &= \alpha(3^-) - \alpha(2^+) = 5 - 4 = 1,\\
\mu_\alpha([2,2]) &= \alpha(2^+) - \alpha(2^-) = 4 - 2 = 2,\\
\mu_\alpha((-1,3)) &= \alpha(3^-) - \alpha(-1^+) = 5 - 0 = 5,\\
\mu_\alpha\left(\left[-8,\tfrac{1}{2}\right]\right) &= \alpha\left(\tfrac{1}{2}^+\right) - \alpha(-8^-) = 0 - 0 = 0.
\end{aligned}$$

It can be seen from these examples that the α-measure of an interval takes account of a jump in the value of α at an endpoint if and only if that endpoint is included in the interval. Note also that it is the left- and right-hand *limits* of α at the endpoints that determine the measure, *not the value* of α at the endpoints. Note finally that, as the following examples illustrate, an interval that has one or both endpoints infinite may have, but does not necessarily have, infinite measure:

$$\begin{aligned}
\mu_\alpha([2,\infty)) &= \alpha(\infty^-) - \alpha(2^-) = \infty - 2 = \infty,\\
\mu_\alpha((-\infty,\infty)) &= \alpha(\infty^-) - \alpha((-\infty)^+) = \infty - 0 = \infty,\\
\mu_\alpha((-\infty,2]) &= \alpha(2^+) - \alpha((-\infty)^+) = 4 - 0 = 4.
\end{aligned}$$

Exercises 4-1:

1. Let $\alpha : \mathbb{R} \to \mathbb{R}$ be defined by

$$\alpha(x) = \begin{cases} x, & \text{if } x < 0, \\ 1, & \text{if } x = 0, \\ 3 - e^{-x}, & \text{if } x > 0. \end{cases}$$

   (a) Sketch the graph of α.

   (b) Find $\mu_\alpha((0,1))$, $\mu_\alpha([0,1])$, $\mu_\alpha((-1,1))$, $\mu_\alpha([0,0])$, $\mu_\alpha((-\infty,1))$, $\mu_\alpha((0,\infty)$, $\mu_\alpha([0,\infty))$.

2. Let $\alpha : \mathbb{R} \to \mathbb{R}$ be defined by

$$\alpha(x) = \begin{cases} 0, & \text{if } x \le 0, \\ 1, & \text{if } 0 < x < 1, \\ 4, & \text{if } 1 \le x < 2, \\ 6, & \text{if } x \ge 2. \end{cases}$$

   (a) Sketch the graph of α.

   (b) Find $\mu_\alpha([-1,2))$, $\mu_\alpha((1,\infty))$, $\mu_\alpha((-\infty,4))$, $\mu_\alpha((0,2])$, $\mu_\alpha((1/2,3/2))$, $\mu_\alpha([1,3])$, $\mu_\alpha((1,3))$.

4.2 Probability Measures

A particularly important type of measure arises when the function α is a *probability distribution function*. In this case, the variable x is referred to as a *random variable*, and for each real number X, the value $\alpha(X)$ is the probability that the random variable x has a value no greater than X:

$$\alpha(X) = P(x \le X).$$

The corresponding α-measure is then called a **probability measure**, and has the property that for any interval I,

$$\mu_\alpha(I) = P(x \in I).$$

Any probability distribution function must necessarily satisfy the conditions $\alpha((-\infty)^+) = 0$ and $\alpha(\infty^-) = 1$, and it follows from this

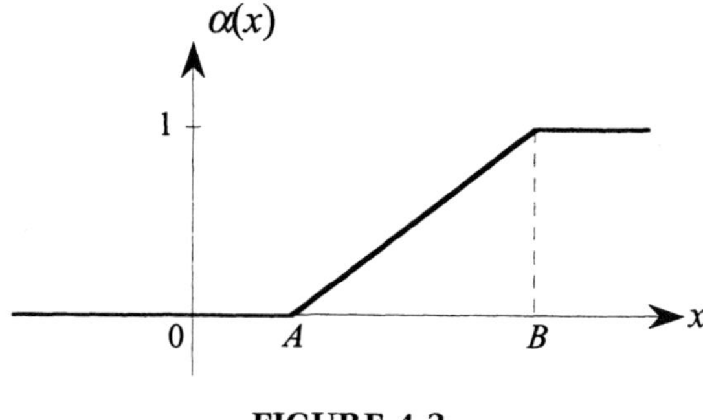

FIGURE 4.2

that if μ_α is a probability measure, then $\mu_\alpha(I) \leq 1$ for any interval I.

Example 4-2-1:
The **uniform distribution** on the interval $[A, B]$ (A and B finite, $A < B$) is the probability distribution α defined by

$$\alpha(x) = \begin{cases} 0, & \text{if } x < A, \\ \frac{x-A}{B-A}, & \text{if } A \leq x \leq B, \\ 1, & \text{if } x > B \end{cases}$$

(cf. Figure 4.2). Since α is continuous, we can say that if I is an interval with finite endpoints a, b, then $\mu_\alpha(I) = \alpha(b) - \alpha(a)$, so that if $A \leq a \leq b \leq B$, then

$$\mu_\alpha(I) = \frac{b-A}{B-A} - \frac{a-A}{B-A} = \frac{b-a}{B-A} = \frac{\text{length of } I}{\text{length of } [A, B]}.$$

Since the only changes in the value of α occur within the interval $[A, B]$, it follows that for any interval I,

$$\mu_\alpha(I) = \frac{\text{length of } I \cap [A, B]}{\text{length of } [A, B]}.$$

In this case $\mu_\alpha(I)$ can be interpreted as the probability that a random number generator, programmed to select a random number in the interval $[A, B]$, will in fact select a number in I.

Example 4-2-2:
A **discrete distribution** is a probability distribution that is constant except for jump discontinuities at a finite or countably infinite

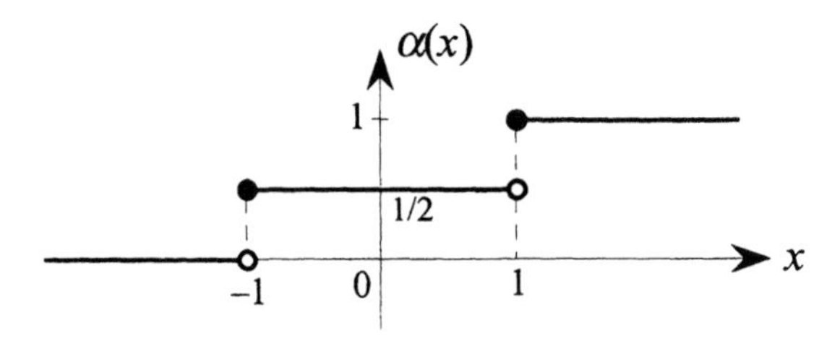

FIGURE 4.3

number of points. An example is the function α defined by

$$\alpha(x) = \begin{cases} 0, & \text{if } x < -1, \\ \frac{1}{2}, & \text{if } -1 \leq x < 1, \\ 1, & \text{if } x \geq 1 \end{cases}$$

(cf. Figure 4.3). In this case we have

$$\mu_\alpha(x) = \begin{cases} 0, & \text{if } I \text{ contains neither 1 nor } -1, \\ \frac{1}{2}, & \text{if } I \text{ contains 1 or } -1 \text{ but not both}, \\ 1, & \text{if } I \text{ contains both 1 and } -1. \end{cases}$$

This corresponds to a random variable x such that

$$P(x = -1) = P(x = 1) = \frac{1}{2}.$$

For example, x might be the outcome of tossing a coin if "heads" is scored as 1 and "tails" as -1.

Exercises 4-2:

1. If x is a random variable that can take only one value λ (with probability 1), what is the corresponding probability distribution function α?

2. If x is a random variable that can take exactly n values $\lambda_1, \lambda_2, \ldots, \lambda_n$ (where $\lambda_1 < \lambda_2 < \cdots < \lambda_n$), each with a probability $1/n$, what is the corresponding probability distribution function α?

4.3 Simple Sets

A **simple set** is a subset of $\mathbb{R}$ that can be expressed as the union of a *finite* collection of disjoint intervals. If S is a simple set, say $S = \cup_{j=1}^{m} I_j$ where $I_1, I_2, \ldots, I_m$ are disjoint intervals, and if $\alpha : \mathbb{R} \to \mathbb{R}$ is a monotone increasing function, then the α-measure of S is defined by

$$\mu_\alpha(S) = \sum_{j=1}^{m} \mu_\alpha(I_j).$$

A given simple set can, of course, be subdivided into disjoint intervals in many different ways, but the value of its α-measure is independent of the way in which it is subdivided. Note also that

(i) $\mu_\alpha(S) \geq 0$ for any simple set S;

(ii) if S and T are simple sets such that $S \subseteq T$, then $\mu_\alpha(S) \leq \mu_\alpha(T)$.

Some other elementary properties of simple sets are explored in the exercises. Note finally that a simple set is said to be **α-finite** if it has finite α-measure.

Exercises 4-3:

1. It is true (though rather tedious and unenlightening to prove in general) that if S and T are simple sets, so are $S \cup T$, $S \cap T$, and $S - T = \{x : x \in S \text{ and } x \notin T\}$. Verify this for each of the following cases:

   (a) $S = [1, 3) \cup (4, 8)$, $T = (2, 5] \cup (6, 7]$;

   (b) $S = (2, 3) \cup [5, 7]$, $T = [1, 4] \cup [6, 8)$;

   (c) $S = (1, 2] \cup [5, 6)$, $T = [2, 4] \cup (5, 7)$.

2. Prove that if S and T are disjoint simple sets, then $\mu_\alpha(S \cup T) = \mu_\alpha(S) + \mu_\alpha(T)$ for any monotone increasing function $\alpha : \mathbb{R} \to \mathbb{R}$. Give examples to show that if S and T are not disjoint, then $\mu_\alpha(S \cup T)$ may or may not equal $\mu_\alpha(S) + \mu_\alpha(T)$.

3. Use what was proved in the preceding exercise to show that if S and T are simple sets such that $T \subseteq S$ and T is α-finite, then $\mu_\alpha(S - T) = \mu_\alpha(S) - \mu_\alpha(T)$ for any monotone increasing function $\alpha : \mathbb{R} \to \mathbb{R}$. (Note that T is required to be α-finite in order to avoid

having the meaningless expression $\infty - \infty$ on the right-hand side.)

4. Give examples to show that if T is not a subset of S, then $\mu_\alpha(S-T)$ may or may not equal $\mu_\alpha(S) - \mu_\alpha(T)$.

4.4 Step Functions Revisited

Let $\alpha : \mathbb{R} \to \mathbb{R}$ be a monotone increasing function. Let I be any interval, and let $\theta : I \to \mathbb{R}$ be a step function. It is clear from the relevant definitions that the support of θ is a simple set. We say that θ is **α-summable** if the support of θ is α-finite. In that case we associate with θ a real number $A_\alpha(\theta)$ defined by

$$A_\alpha(\theta) = \sum_{j=1}^{n} c_j \mu_\alpha(I_j)$$

using the notation introduced in Section 2-5.

If $\alpha(x) = x$ for all $x \in \mathbb{R}$, so that $\mu_\alpha(I_j)$ is just the ordinary length of the interval I_j, then $A_\alpha(\theta)$ is just the area $A(\theta)$ under the graph of θ, as defined in Section 2.5. In general, $A_\alpha(\theta)$ can be thought of as a generalized "area" for which "lengths" along the x-axis are measured by α-measure rather than by ordinary length.

Note that if the endpoints of I are both finite, then any step function $\theta : I \to \mathbb{R}$ is α-summable for all monotone increasing functions $\alpha : \mathbb{R} \to \mathbb{R}$.

Example 4-4-1:
Let α be defined as in Example 4-1-1, and let $\theta_1 : [0,3] \to \mathbb{R}$ be defined by

$$\theta_1(x) = \begin{cases} -1, & \text{if } 0 \le x < 1, \\ 2, & \text{if } 1 \le x \le 3 \end{cases}$$

(cf. Figure 4.4). Then

$$\mu_\alpha([0,1)) = \alpha(1^-) - \alpha(0^-) = 0 - 0 = 0,$$
$$\mu_\alpha([1,3]) = \alpha(3^+) - \alpha(1^-) = 5 - 0 = 5,$$

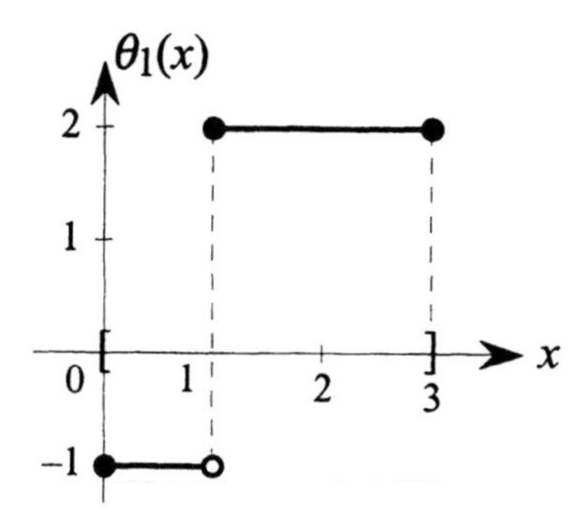

FIGURE 4.4

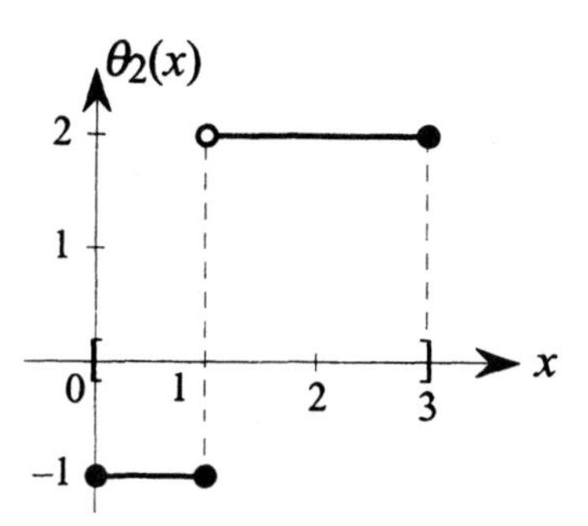

FIGURE 4.5

and so

$$A_\alpha(\theta_1) = (-1)0 + 2(5) = 10.$$

Suppose we modify the definition of θ_1 very slightly, to give

$$\theta_2(x) = \begin{cases} -1, & \text{if } 0 \le x \le 1, \\ 2, & \text{if } 1 < x \le 3 \end{cases}$$

(cf. Figure 4.5). In this case

$$\mu_\alpha([0,1]) = \alpha(1^+) - \alpha(0^-) = 1 - 0 = 1,$$
$$\mu_\alpha((1,3]) = \alpha(3^+) - \alpha(1^+) = 5 - 1 = 4,$$

and so

$$A_\alpha(\theta_2) = (-1)1 + 2(4) = 7.$$

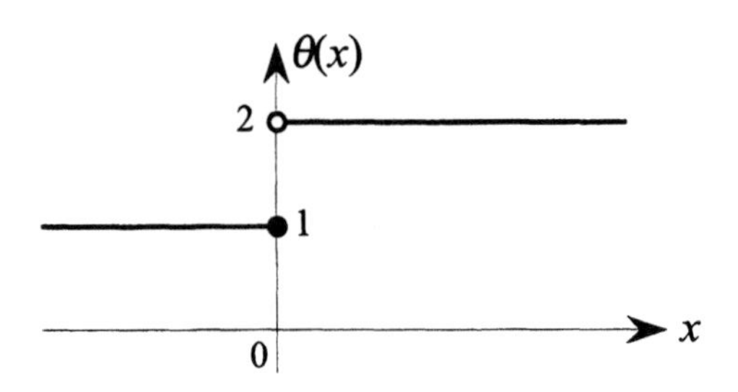

FIGURE 4.6

Note that while the area $A(\theta)$ under the graph is the same for these two functions, the values of $A_\alpha(\theta_1)$ and $A_\alpha(\theta_2)$ are different. This is because at the single point where θ_1 and θ_2 have different values, α has a discontinuity, and so the interval consisting of that single point has positive α-measure. Clearly, discontinuities in α complicate matters!

Example 4-4-2:
Let α be the discrete distribution function defined in Example 4-2-2, and let $\theta : \mathbb{R} \to \mathbb{R}$ be defined by

$$\theta(x) = \begin{cases} 1, & \text{if } x \leq 0, \\ 2, & \text{if } x > 0 \end{cases}$$

(cf. Figure 4.6). Then

$$\mu_\alpha((-\infty, 0]) = \alpha(0^+) - \alpha((-\infty)^+) = \frac{1}{2} - 0 = \frac{1}{2},$$
$$\mu_\alpha((0, \infty)) = \alpha(\infty^-) - \alpha(0^+) = 1 - \frac{1}{2} = \frac{1}{2},$$

and so

$$A_\alpha(\theta) = 1\left(\frac{1}{2}\right) + 2\left(\frac{1}{2}\right) = \frac{3}{2}.$$

We conclude this section by listing a number of basic properties that are straightforward to prove and intuitively reasonable, so we will omit the proofs.

(i) If θ is a nonnegative α-summable step function, then $A_\alpha(\theta) \geq 0$; also, $A_\alpha(0) = 0$.

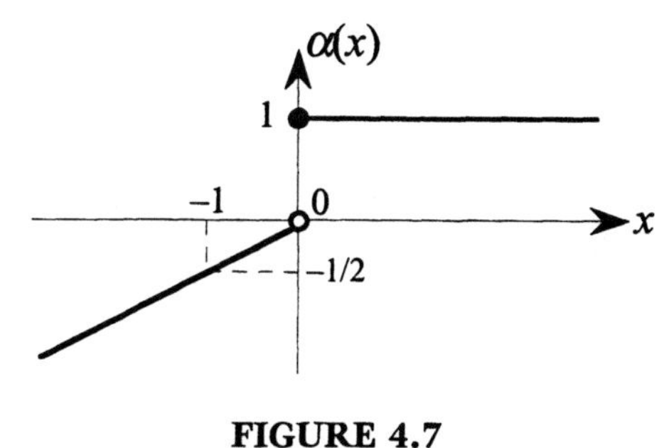

FIGURE 4.7

(ii) If θ_1 and θ_2 are α-summable step functions on the same interval I such that $\theta_1 \leq \theta_2$ on I, then $A_\alpha(\theta_1) \leq A_\alpha(\theta_2)$.

(iii) If θ is an α-summable step function, then so are $|\theta|$, θ^+, and θ^-, and we have $A_\alpha(\theta) = A_\alpha(\theta^+) + A_\alpha(\theta^-)$ and $A_\alpha(|\theta|) = A_\alpha(\theta^+) - A_\alpha(\theta^-)$.

(iv) If $\theta_1, \theta_2, \ldots, \theta_m$ are α-summable step functions on the same interval I, $a_1, a_2, \ldots, a_m$ are finite real numbers, and $\theta : I \to \mathbb{R}$ is defined by

$$\theta(x) = \sum_{j=1}^{m} a_j \theta_j(x)$$

for all $x \in I$ (i.e., $\theta = \sum_{j=1}^{m} a_j \theta_j$), then θ is also an α-summable step function on I, and

$$A_\alpha(\theta) = \sum_{j=1}^{m} a_j A_\alpha(\theta_j).$$

Exercises 4-4:
Let $\alpha : \mathbb{R} \to \mathbb{R}$ be defined by

$$\alpha(x) = \begin{cases} \frac{1}{2}x, & \text{if } x < 0, \\ 1, & \text{if } x \geq 0 \end{cases}$$

(cf. Figure 4.7). For each of the following step functions θ:

(a) sketch the graph of θ;

(b) determine whether or not θ is α-summable, and if it is, find $A_\alpha(\theta)$.

1. $\theta : (-2, 1) \to \mathbb{R}$ defined by

$$\theta(x) = \begin{cases} 3, & \text{if } -2 < x < -1, \\ -1, & \text{if } -1 \le x < 1. \end{cases}$$

2. $\theta : [-1, 1] \to \mathbb{R}$ defined by

$$\theta(x) = \begin{cases} -2, & \text{if } -1 \le x < 0, \\ 1, & \text{if } 0 \le x \le 1. \end{cases}$$

3. $\theta : [-1, \infty) \to \mathbb{R}$ defined by

$$\theta(x) = \begin{cases} 2, & \text{if } -1 \le x \le 3, \\ 1, & \text{if } x > 3. \end{cases}$$

4. $\theta : (-\infty, 0) \to \mathbb{R}$ defined by

$$\theta(x) = \begin{cases} 0, & \text{if } x < -1, \\ 1, & \text{if } -1 \le x < 0. \end{cases}$$

5. $\theta : \mathbb{R} \to \mathbb{R}$ defined by

$$\theta(x) = \begin{cases} -1, & \text{if } x < 0, \\ 1, & \text{if } x \ge 0. \end{cases}$$

4.5 Definition of the Integral

We are now in a position to set up the necessary machinery for defining the Lebesgue–Stieltjes integral. Throughout this section we take I to be a given interval with endpoints a, b, and $\alpha : \mathbb{R} \to \mathbb{R}$ to be a given monotone increasing function.

Let $f : I \to \mathbb{R}$ be a function that is *nonnegative* on I. A sequence $\theta_1, \theta_2, \theta_3, \ldots$ is said to be **admissible for** f if it satisfies all the following conditions:

(a) θ_j is an α-summable step function on I, for each $j = 1, 2, 3, \ldots$;

(b) $\theta_j(x) \ge 0$ for each $x \in I$ and each $j = 1, 2, 3, \ldots$;

(c) $0 \le f(x) \le \sum_{j=1}^{\infty} \theta_j(x)$ for each $x \in I$.

Theorem 4.5.1
An admissible sequence exists for any nonnegative function $f : I \to \mathbb{R}$.

Proof There are two cases to consider.

Case 1. The endpoints a, b of I are finite. In this case we define the function $\theta_j : I \to \mathbb{R}$ by $\theta_j(x) = 1$ for all $x \in I$ ($j = 1, 2, 3 \ldots$). Then θ_j is nonnegative, and since the endpoints of I are finite, θ_j is certainly α-summable for each $j = 1, 2, 3 \ldots$. Since $\sum_{j=1}^{\infty} \theta_j(x) = \sum_{j=1}^{\infty} 1 = \infty$, condition (c) above is also satisfied for any nonnegative function $f : I \to \mathbb{R}$, and so the sequence $\theta_1, \theta_2, \theta_3, \ldots$ is admissible for f.

Case 2. Either $a = -\infty$ or $b = \infty$ or both. In this case we define the subinterval I_j of I as follows:

If I is $(-\infty, b)$, b finite, then I_j is $(b - j, b)$.
If I is $(-\infty, b]$, b finite, then I_j is $(b - j, b]$.
If I is (a, ∞), a finite, then I_j is $(a, a + j)$.
If I is $[a, \infty)$, a finite, then I_j is $[a, a + j)$.
If I is $(-\infty, \infty)$, then I_j is $(-j, j)$.

For each $j = 1, 2, 3, \ldots$ we then define $\theta_j : I \to \mathbb{R}$ by

$$\theta_j(x) = \begin{cases} 1, & \text{if } x \in I_j, \\ 0, & \text{if } x \in I - I_j. \end{cases}$$

Then θ_j is nonnegative on I and is α-summable, since its support I_j is an interval with finite endpoints. Further, for each $x \in I$ we have that $\theta_j(x) = 1$ for all sufficiently large values of j, and so again $\sum_{j=1}^{\infty} \theta_j(x) = \infty$ for each $x \in I$. Thus the sequence $\theta_1, \theta_2, \theta_3, \ldots$ is admissible for any nonnegative function $f : I \to \mathbb{R}$. □

We associate with any *nonnegative* function $f : I \to \mathbb{R}$ an extended real number $L_\alpha(f)$ defined by

$$L_\alpha(f) = \inf \left\{ \sum_{j=1}^{\infty} A_\alpha(\theta_j) \right\},$$

where the greatest lower bound is taken over all sequences $\theta_1, \theta_2, \theta_3, \ldots$ that are admissible for f. Since the set of all such sums $\sum_{j=1}^{\infty} A_\alpha(\theta_j)$ is non-empty (by Theorem 4.5.1) and has 0 as a lower bound, it follows by Theorem 1.3.1 that $L_\alpha(f)$ exists and $L_\alpha(f) \geq 0$ for any nonnegative function $f : I \to \mathbb{R}$. Note that $L_\alpha(f) = \infty$ is a possibility. Note also that it follows easily from the definition that $L_\alpha(0) = 0$.

Example 4-5-1:
Let $f : [0, 1] \to \mathbb{R}$ be the function discussed in Section 3.3, and let the functions θ_j be as defined in that section. For each $j = 1, 2, 3, \ldots, \theta_j$ is

a nonnegative α-summable step function on $[0, 1]$, and $\sum_{j=1}^{\infty} \theta_j(x) = f(x)$ for each $x \in I$, so the sequence $\theta_1, \theta_2, \theta_3, \ldots$ is admissible for f. In particular, let α^* be defined by $\alpha^*(x) = x$ for all $x \in \mathbb{R}$. Then for each $j = 1, 2, 3, \ldots$ we have

$$A_{\alpha^*}(\theta_j) = \mu_{\alpha^*}([r_j, r_j]) = r_j - r_j = 0,$$

and so $\sum_{j=1}^{\infty} A_{\alpha^*}(\theta_j) = 0$. It follows that $L_{\alpha^*}(f) = 0$.

Note that although each θ_j is a step function, $f = \sum_{j=1}^{\infty} \theta_j$ is not a step function, since it is not possible to describe it by taking constant values on any *finite* set of subintervals of $[0, 1]$.

Theorem 4.5.2

For any function $f : I \to \mathbb{R}$ we have:

(i) $L_\alpha(f^+) \leq L_\alpha(|f|)$ and $L_\alpha(-f^-) \leq L_\alpha(|f|)$;

(ii) $L_\alpha(|af|) = |a| L_\alpha(|f|)$ for any finite nonzero real number a (and for $a = 0$, provided that $L_\alpha(|f|)$ is finite).

Proof (i) Clearly, any sequence $\theta_1, \theta_2, \theta_3, \ldots$ that is admissible for $|f|$ is also admissible for f^+ and $-f^-$. Thus the set of admissible sequences for $|f|$ is a subset of the set of admissible sequences for f^+ and of the set of admissible sequences for $-f^-$. Part (i) follows at once from Exercises 1-3, No. 3. □

(ii) It is obvious that both sides are zero if $a = 0$ and $L_\alpha(|f|)$ is finite, so suppose $a \neq 0$. Let $\theta_1, \theta_2, \theta_3, \ldots$ be an admissible sequence for $|f|$, so that for all $x \in I$,

$$0 \leq |f(x)| \leq \sum_{j=1}^{\infty} \theta_j(x).$$

Then

$$0 \leq |af(x)| \leq |a| \sum_{j=1}^{\infty} \theta_j(x) = \sum_{j=1}^{\infty} |a| \theta_j(x)$$

for all $x \in I$, and also

$$\sum_{j=1}^{\infty} A_\alpha(|a|\theta_j) = \sum_{j=1}^{\infty} |a| A_\alpha(\theta_j) = |a| \sum_{j=1}^{\infty} A_\alpha(\theta_j),$$

and so $|a|\theta_1, |a|\theta_2, |a|\theta_3, \ldots$ is an admissible sequence for $|af|$.

Conversely, if $\theta_1, \theta_2, \theta_3, \ldots$ is an admissible sequence for $|af|$, then $(1/|a|)\theta_1, (1/|a|)\theta_2, (1/|a|)\theta_3, \ldots$ is an admissible sequence for $|f|$. Thus, if S is the set of admissible sequences for $|f|$. Then from above and Exercises 1-3, No. 4(a), we have

$$\begin{aligned} L_\alpha(|af|) &= \inf_S \left\{ \sum_{j=1}^{\infty} A_\alpha(|a|\theta_j) \right\} = \inf_S \left\{ \sum_{j=1}^{\infty} |a| A_\alpha(\theta_j) \right\} \\ &= |a| \inf_S \left\{ \sum_{j=1}^{\infty} A_\alpha(\theta_j) \right\} \\ &= |a| L_\alpha(|f|). \end{aligned}$$

□

Theorem 4.5.3

If $f, g : I \to \mathbb{R}$ are such that $0 \le f(x) \le g(x)$ for all $x \in I$, then $L_\alpha(f) \le L_\alpha(g)$.

Proof This result follows at once, since any sequence $\theta_1, \theta_2, \theta_3, \ldots$ that is admissible for g is also admissible for f. □

Theorem 4.5.4

If $f_1, f_2, f_3, \ldots$ is a sequence of functions such that $f_j : I \to \mathbb{R}$ for each $j = 1, 2, 3, \ldots$ and $\sum_{j=1}^{\infty} f_j(x)$ converges for each $x \in I$, then

$$L_\alpha\left(\left| \sum_{j=1}^{\infty} f_j \right| \right) \le \sum_{j=1}^{\infty} L_\alpha(|f_j|).$$

[Note that is it possible to have $\sum_{j=1}^{\infty} L_\alpha(|f_j|) = \infty$ under the conditions of the theorem. Note also that when the sequence $f_1, f_2, f_3, \ldots$ is such that $f_j = 0$ for all $j > n$, we obtain the important special case $L_\alpha(|\sum_{j=1}^{n} f_j|) \le \sum_{j=1}^{n} L_\alpha(|f_j|)$.]

Proof The result is evidently true if $\sum_{j=1}^{\infty} L_\alpha(|f_j|) = \infty$, so assume that $\sum_{j=1}^{\infty} L_\alpha(|f_j|)$ is finite. Then certainly $L_\alpha(|f_j|)$ is finite for each $j = 1, 2, 3, \ldots$. Take any $\epsilon > 0$. By part (ii) of Theorem 1.3.2 and the definition of $L_\alpha(|f_j|)$, we have that for each $j = 1, 2, 3, \ldots$ there is a sequence $\theta_{j1}, \theta_{j2}, \theta_{j3}, \ldots$ admissible for $|f_j|$ and such that

$$\sum_{m=1}^{\infty} A_\alpha(\theta_{jm}) \le L_\alpha(|f_j|) + 2^{-j}\epsilon.$$

It follows from the properties of double series of positive terms (Section 2-2) that

$$\begin{aligned}\sum_{j,m=1}^{\infty} A_\alpha(\theta_{jm}) &= \sum_{j=1}^{\infty}\left(\sum_{m=1}^{\infty} A_\alpha(\theta_{jm})\right)\\ &\le \sum_{j=1}^{\infty}(L_\alpha(|f_j|) + 2^{-j}\epsilon)\\ &= \sum_{j=1}^{\infty} L_\alpha(|f_j|) + \epsilon\sum_{j=1}^{\infty}\left(\frac{1}{2}\right)^j\\ &= \sum_{j=1}^{\infty} L_\alpha(|f_j|) + \epsilon\left(\frac{1/2}{1-1/2}\right)\\ &= \sum_{j=1}^{\infty} L_\alpha(|f_j|) + \epsilon.\end{aligned}$$

But since $0 \le |f_j(x)| \le \sum_{m=1}^{\infty} \theta_{jm}(x)$ for all $x \in I$ and each $j = 1, 2, 3, \ldots$, we have

$$0 \le \left|\sum_{j=1}^{\infty} f_j(x)\right| \le \sum_{j=1}^{\infty} |f_j(x)| \le \sum_{j=1}^{\infty}\left(\sum_{m=1}^{\infty} \theta_{jm}(x)\right) = \sum_{j,m=1}^{\infty} \theta_{jm}(x).$$

If we write the double sequence θ_{jm} $(j, m = 1, 2, 3, \ldots)$ as a single sequence (in any way we choose), the result is a sequence $\psi_1, \psi_2, \psi_3, \ldots$ that is admissible for $|\sum_{j=1}^{\infty} f_j|$, and so

$$L_\alpha\left(\left|\sum_{j=1}^{\infty}(f_j)\right|\right) \le \sum_{i=1}^{\infty} A_\alpha(\psi_i) = \sum_{j,m=1}^{\infty} A_\alpha(\theta_{jm}) \le \sum_{j=1}^{\infty} L_\alpha(|f_j|) + \epsilon;$$

thus, $L_\alpha(|\sum_{j=1}^{\infty}(f_j)|) - \sum_{j=1}^{\infty} L_\alpha(|f_j|) \le \epsilon$ for any $\epsilon > 0$, and it follows that

$$L_\alpha\left(\left|\sum_{j=1}^{\infty}(f_j)\right|\right) - \sum_{j=1}^{\infty} L_\alpha(|f_j|) \le 0,$$

i.e.,

$$L_\alpha\left(\left|\sum_{j=1}^{\infty}(f_j)\right|\right) \le \sum_{j=1}^{\infty} L_\alpha(|f_j|),$$

as required. □

Theorem 4.5.5
If $f_1, f_2, f_3, \ldots$ is a sequence of functions such that $f_n : I \to \mathbb{R}$ and $L_\alpha(|f_n|)$ is finite for each $n = 1, 2, 3, \ldots$, and if $f : I \to \mathbb{R}$ is such that $L_\alpha(|f - f_n|) \to 0$ (as $n \to \infty$) then:

(i) $L_\alpha(|f|)$, $L_\alpha(f^+)$ *and* $L_\alpha(-f^-)$ *are all finite;*

(ii) $L_\alpha(|f^+ - f_n^+|) \to 0$, $L_\alpha(|f^- - f_n^-|) \to 0$ *and* $L_\alpha(||f| - |f_n||) \to 0$;

(iii) $L_\alpha(|f_n|) \to L_\alpha(|f|)$, $L_\alpha(f_n^+) \to L_\alpha(f^+)$ *and* $L_\alpha(-f_n^-) \to L_\alpha(-f^-)$.

Proof (i) Since $L_\alpha(|f - f_n|) \to 0$, there certainly exists a positive integer N such that $L_\alpha(|f - f_N|)$ is finite. Then

$$L_\alpha(|f|) = L_\alpha(|f_N + f - f_N|) \le L_\alpha(|f_N|) + L_\alpha(|f - f_N|),$$

by Theorem 4.5.4, and so $L_\alpha(|f|)$ is finite, since $L_\alpha(|f_N|)$ and $L_\alpha(|f - f_N|)$ are both finite. The fact that $L_\alpha(f^+)$ and $L_\alpha(-f^-)$ are finite follows at once from Theorem 4.5.2(i).

(ii) From Exercises 2-6 we have that on I,

$$|f^+ - f_n^+| \le |f - f_n|, \; |f^- - f_n^-| \le |f - f_n|,$$

and

$$||f| - |f_n|| \le |f - f_n|.$$

Since $L_\alpha(|f - f_n|) \to 0$ by hypothesis, part (ii) follows at once using Theorem 4.5.3.

(iii) We have, using Theorem 4.5.4:

$$\begin{aligned} L_\alpha(|f_n|) &= L_\alpha(|f + f_n - f|) \le L_\alpha(|f|) + L_\alpha(|f_n - f|) \\ &= L_\alpha(|f|) + L_\alpha(|f - f_n|). \end{aligned} \tag{4.1}$$

Also,

$$L_\alpha(|f|) = L_\alpha(|f - f_n + f_n|) \le L_\alpha(|f - f_n|) + L_\alpha(|f_n|),$$

and so

$$L_\alpha(|f_n|) \ge L_\alpha(|f|) - L_\alpha(|f - f_n|). \tag{4.2}$$

From equations (4.1) and (4.2) we have

$$L_\alpha(|f|) - L_\alpha(|f - f_n|) \le L_\alpha(|f_n|) \le L_\alpha(|f|) + L_\alpha(|f - f_n|),$$

i.e.,

$$-L_\alpha(|f - f_n|) \le L_\alpha(|f_n|) - L_\alpha(|f|) \le L_\alpha(|f - f_n|),$$

i.e.,

$$|L_\alpha(|f_n|) - L_\alpha(|f|)| \le L_\alpha(|f - f_n|).$$

Since $L_\alpha(|f - f_n|) \to 0$ by hypothesis, it follows that $L_\alpha(|f_n|) \to L_\alpha(|f|)$. We have also

$$f_n^+ = f^+ + (f_n^+ - f^+) \quad \text{and} \quad f^+ = (f^+ - f_n^+) + f_n^+,$$
$$-f_n^- = -f^- + (-f_n^- - (-f^-)) \quad \text{and} \quad -f^- = (-f^- - (-f_n^-)) + (-f_n^-),$$

and since $L_\alpha(|f^+ - f_n^+|) \to 0$ and $L_\alpha(|-f^- - (-f_n^-)|) \to 0$ by part (ii), the rest of part (iii) can be proved by an argument similar to the preceding one. □

Now let $\alpha : \mathbb{R} \to \mathbb{R}$ be a monotone increasing function, and let $f : I \to \mathbb{R}$ be a function with the property that there is a sequence $\theta_1, \theta_2, \theta_3, \dots$ of α-summable step functions defined on I such that $L_\alpha(|f - \theta_n|) \to 0$. It follows from part (i) of Theorem 4.5.5 that $L_\alpha(|f|)$, $L_\alpha(f^+)$ and $L_\alpha(-f^-)$ are all finite. Under these conditions we write

$$\int_I f\, d\alpha = L_\alpha(f^+) - L_\alpha(-f^-)$$

and call this quantity the (Lebesgue–Stieltjes) **integral** of f over I with respect to α. If $\alpha(x) = x$ for all $x \in \mathbb{R}$, we write $\int_I f\, dx$ instead of $\int_I f\, d\alpha$; this special case is called the **Lebesgue integral** of f over I.

It is important to be sure that $\int_I \theta\, d\alpha = A_\alpha(\theta)$ for any α-summable step function $\theta : I \to \mathbb{R}$, because only then can the integral justifiably be regarded as an extension of the concept of the "area under a graph" as defined for step functions. This is, in fact, true, but the proof is surprisingly hard, so we shall just state the result without proof:

Theorem 4.5.6
For any α-summable step function $\theta : I \to \mathbb{R}$, we have $\int_I \theta\, d\alpha = A_\alpha(\theta)$.

Exercises 4-5:

1. Complete the proof of part (iii) of Theorem 4.5.5 by showing that $L_\alpha(f_n^+) \to L_\alpha(f^+)$ and $L_\alpha(-f_n^-) \to L_\alpha(-f^-)$.
2. Let f be the function discussed in Section 3.3 and Example 4-5-1, and let α^* be defined by $\alpha^*(x) = x$ for all $x \in \mathbb{R}$. Since $f^+ = f$ and $f^- = 0$, and we have shown in Example 4-5-1 that $L_{\alpha^*}(f) = 0$,

it follows that $L_{\alpha^*}(f^+) - L_{\alpha^*}(-f^-) = 0$. Complete the proof of the fact that $\int_{[0,1]} f\,dx = 0$, by showing that there is a sequence $\theta_1, \theta_2, \theta_3, \ldots$ of α^*-summable step functions on $[0,1]$ such that $L_{\alpha^*}(|f - \theta_n|) \to 0$. (This example shows that $\int_I f\,dx$ may exist in the Lebesgue–Stieltjes sense in some cases where f is not Riemann integrable over I).

3. Use Theorem 4.5.6 to show that if $\theta : I \to \mathbb{R}$ is a nonnegative α-summable step function, then $A_\alpha(\theta) = L_\alpha(\theta)$.

4.6 The Lebesgue Integral

The mathematical machinery required to define the Lebesgue–Stieltjes integral is notably more complex than that needed for the Riemann integral. Here, we pause to discuss informally the definition of the Lebesgue integral and contrast it with that of the Riemann integral.

Let I be an interval and $f : I \to \mathbb{R}$ be some function, which for simplicity we assume to be nonnegative. If f is Lebesgue integrable over I, then $L_x(f) < \infty$ and

$$\int_I f\,dx = L_x(f).$$

Recall that

$$L_x(f) = \inf \sum_{j=0}^{\infty} A_j(\theta_j),$$

where the greatest lower bound is taken over all the sequences $\{\theta_j\}$ that are admissible for f. If $\{\theta_j\}$ is an admissible sequence, then in particular, $0 \le f(x) \le \sum_{j=1}^{\infty} \theta_j(x)$. For a given admissible sequence $\{\theta_j\}$ the quantity $\sum_{j=0}^{\infty} A(\theta_j)$ is thus an upper bound for the "area under the graph of f," and the quantity $L_x(f)$ plays a role in the Lebesgue theory analogous to that of the upper Riemann–Darboux integral in the Riemann theory. Suppose now we consider sequences $\{\phi_j\}$ that satisfy conditions (a) and (b) for an admissible sequence (see p. 60), but satisfy the inequality (c*) $0 \leq \sum_{j=1}^{\infty} A(\phi_j) \leq f(x)$

for all $x \in I$ instead of condition (c). Let

$$\ell_x(f) = \sup \sum_{j=1}^{\infty} A(\phi_j),$$

where the least upper bound is taken over all sequences satisfying conditions (a), (b), and (c*). If $\ell_x(f) < \infty$, we could define another integral by

$$\underline{\int_I} f \, dx = \ell_x(f).$$

The quantity $\ell_x(f)$ is analogous to the lower Riemann–Darboux integral. Note that in the Riemann theory the upper and lower Riemann–Darboux integrals are always finite even if f is not Riemann integrable. This is because the definition of Riemann integrability is framed in terms of closed intervals on which f must be bounded. In the Lebesgue theory, no restrictions are placed on I, and f need not be bounded; consequently, neither $\ell_x(f)$ nor $L_x(f)$ need be finite.

If f is Riemann integrable, then condition (ii) of the definition requires that the upper and lower Riemann–Darboux integrals be equal. There is no analogue to this condition in the Lebesgue theory: It is not required that $\ell_x(f) = L_x(f)$ for the Lebesgue integral to exist, only that $L_x(f) < \infty$. Prima facie, this may seem a weakness in the theory, as there is no particular reason to choose $L_x(f)$ over $\ell_x(f)$ to define an integral, but it can be shown that the relationship $\ell_x(f) = L_x(f)$ is in fact a consequence of the condition $L_x(f) < \infty$ together with measurability of f (see p. 82). In other words, if $L_x(f) < \infty$ and f is measurable, then $\ell_x(f) < \infty$ and $\ell_x(f) = L_x(f)$. A similar statement can be made if $\ell_x(f) < \infty$, and in this sense ℓ_x and L_x are always on the same mathematical "footing" in the Lebesgue theory.

The definitions of the Riemann and Lebesgue integrals differ fundamentally in the functions used to approximate the "area under the graph," and it is this difference that affords greater generality for the Lebesgue integral. Recall that a function f is Riemann integrable on a closed interval I if for any $\epsilon > 0$ there are step functions g^{ϵ}, G^{ϵ} such that $g^{\epsilon} \leq f \leq G^{\epsilon}$ and $A(G^{\epsilon}) - A(g^{\epsilon}) \leq \epsilon$. The definition of the

Lebesgue integral also uses step functions but in a different way. For example, if $\{\theta_j\}$ is an admissible sequence, then the inequality in the Lebesgue definition analogous to the inequality $f \leq G^\epsilon$ in the Riemann definition is $f \leq \sum_{j=1}^{\infty} \theta_j$. Although each θ_j is a step function, the sum $\sum_{j=1}^{\infty} \theta_j$ need not be a step function. A more general class of functions is thus allowed into the "approximation." Coupled with this increased generality is the notion of measure, which is in itself a generalization of length. The generality is enough to make functions such as that described in Section 3.3 integrable under the Lebesgue definition (cf. Exercises 4-5, No. 2). The real value of the generalization, however, lies in results such as the dominated convergence theorem (Theorem 5.3.3), which resolve some of the problems with the Riemann integral discussed at the end of Section 3.3.

The Lebesgue integral is a generalization of the Riemann integral. We state the following result without proof:

Theorem 4.6.1
If a function $f : [a, b] \to \mathbb{R}$ is Riemann integrable over the interval $I = [a, b]$, then it is also Lebesgue integrable over I, and the two integrals are equal.

The above result is used frequently to calculate Lebesgue integrals. Note that the generalization breaks down if the integral is improper. For example, the conditionally convergent integral in Example 3-2-5 is not Lebesgue integrable. (In fact, Theorem 5.1.1 in the next chapter indicates that if a Lebesgue integral exists, it is always absolutely convergent.) If an improper integral is absolutely convergent, however, then it can be shown that it is also Lebesgue integrable and the integrals are equal.

Exercise 4-6:

Let $f : [0, \infty) \to \mathbb{R}$ be defined by

$$f(x) = \begin{cases} \frac{1}{n}, & \text{if } n-1 \leq x < n - \frac{1}{2}, \quad n = 1, 2, \ldots, \\ -\frac{1}{n}, & \text{if } n - \frac{1}{2} \leq x < n, \quad n = 1, 2, \ldots \end{cases}$$

(Figure 4.8).

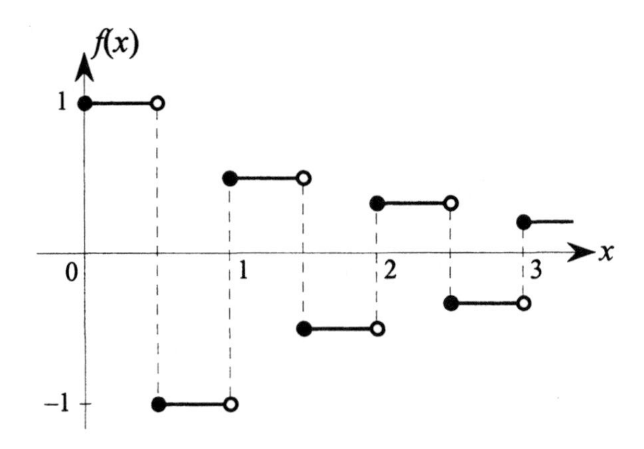

FIGURE 4.8

(a) Evaluate $\int_{[0,c]} f\, dx$ for c a positive real number. (Since f is a step function on $[0, c]$, this integral exists in both the Riemann and Lebesgue senses.)

(b) Show that $\int_0^\infty f\, dx$ exists in the Riemann sense, as $\lim_{c\to\infty} \int_0^c f dx$.

(c) Show that $\lim_{c\to\infty} \int_0^c |f|\, dx$ does not exist; this shows that $\int_{[0,\infty)} f\, dx$ does not exist in the Lebesgue sense (cf. Theorem 5.1.1).

5 CHAPTER Properties of the Integral

In this chapter we will examine some of the essential properties of the Lebesgue–Stieltjes integral, culminating in the convergence theorems that (as remarked in Chapter 3) are among the most important features of the Lebesgue–Stieltjes theory.

5.1 Basic Properties

Theorem 5.1.1
If $f : I \to \mathbb{R}$ is integrable over I with respect to α, then so are f^+, f^-, and $|f|$, and we have

$$\int_I f\, d\alpha = \int_I f^+\, d\alpha + \int_I f^-\, d\alpha \quad \text{and} \quad \int_I |f|\, d\alpha = \int_I f^+\, d\alpha - \int_I f^-\, d\alpha.$$

Proof Since f is integrable over I with respect to α, there is a sequence $\theta_1, \theta_2, \theta_3, \ldots$ of α-summable functions on I such that $L_\alpha(|f - \theta_n|) \to 0$. Since θ_n is α-summable for each $j = 1, 2, 3, \ldots$, so is $|\theta_n|$, and it follows from Exercises 4-5, No. 3, that $L_\alpha(|\theta_n|)$ is finite for each $j = 1, 2, 3, \ldots$. It then follows from part (ii) of Theorem 4.5.5 that

$L_\alpha(|f^+ - \theta_n^+|) \to 0$, $L_\alpha(|f^- - \theta_n^-|) \to 0$ and $L_\alpha(||f| - |\theta_n||) \to 0$, so by definition f^+, f^-, and $|f|$ are all integrable over I with respect to α.

By definition, $\int_I f^+ \, d\alpha = L_\alpha(f^+)$ and $\int_I f^- \, d\alpha = -L_\alpha(-f^-)$, and so

$$\int_I f \, d\alpha = L_\alpha(f^+) - L_\alpha(-f^-) = \int_I f^+ \, d\alpha + \int_I f^- \, d\alpha.$$

Also, we have by property (iii) (at the end of Section 5-4) that

$$A_\alpha(|\theta_n|) = A_\alpha(\theta_n^+) - A_\alpha(\theta_n^-) = A_\alpha(\theta_n^+) + A_\alpha(-\theta_n^-).$$

Thus $L_\alpha(|\theta_n|) = L_\alpha(\theta_n^+) + L_\alpha(-\theta_n^-)$ by Exercises 4-5, No. 3. Letting n tend to ∞, we have by Theorem 4.5.5 part (iii) that

$$L_\alpha(|f|) = L_\alpha(f^+) + L_\alpha(-f^-) = \int_I f^+ \, d\alpha - \int_I f^- \, d\alpha.$$

But since $|f|^+ = |f|$, we have $\int_I |f| \, d\alpha = L_\alpha(|f|)$, which proves the last part of the theorem. □

Theorem 5.1.2

If for each $n = 1, 2, 3, \ldots$, $f_n : I \to \mathbb{R}$ is integrable over I with respect to α, and if $f : I \to \mathbb{R}$ is such that $L_\alpha(|f - f_n|) \to 0$, then f is integrable over I with respect to α, and we have

$$\int_I f_n \, d\alpha \to \int_I f \, d\alpha, \qquad \int_I f_n^+ \, d\alpha \to \int_I f^+ \, d\alpha,$$
$$\int_I f_n^- \, d\alpha \to \int_I f^- \, d\alpha, \qquad \int_I |f_n| \, d\alpha \to \int_I |f| \, d\alpha.$$

Proof Take any positive integer n. We know that f_n is integrable, so by definition there is a sequence $\psi_1, \psi_2, \ldots$ of α-summable step functions such that $L_\alpha(|f_n - \psi_j|) \to 0$ as $j \to \infty$. Thus, for some $k = 1, 2, 3, \ldots$ we must have $L_\alpha(|f_n - \psi_k|) < 1/n$. Choose such a k and denote ψ_k by θ_n. In this way we obtain a sequence $\theta_1, \theta_2, \ldots$ of α-summable step functions such that

$$L_\alpha(|f_n - \theta_n|) < \frac{1}{n}, \tag{5.1}$$

for each $n = 1, 2, \ldots$. Now take any $\epsilon > 0$. Since $L_\alpha(|f - f_n|) \to 0$, there exists a positive integer $N(\epsilon)$ such that

$$n \geq N(\epsilon) \Rightarrow L_\alpha(|f - f_n|) < \frac{\epsilon}{2}. \tag{5.2}$$

Also, from equation (5.1), we have that

$$n \geq \frac{2}{\epsilon} \Rightarrow \frac{1}{n} \leq \frac{\epsilon}{2} \Rightarrow L_\alpha(|f_n - \theta_n|) < \frac{\epsilon}{2}. \tag{5.3}$$

By Theorem 4.5.4 we have that for each $n = 1, 2, \ldots,$

$$L_\alpha(|f - \theta_n|) = L_\alpha(|f - f_n + f_n - \theta_n|) \leq L_\alpha(|f - f_n|) + L_\alpha(|f_n - \theta_n|).$$

Thus, if $n \geq \max\{N(\epsilon), 2/\epsilon\}$ then equations (5.2) and (5.3) imply that

$$L_\alpha(|f - \theta_n|) < \frac{\epsilon}{2} + \frac{\epsilon}{2} = \epsilon,$$

and so $L_\alpha(|f - \theta_n|) \to 0$, and f is integrable over I with respect to α by definition.

The rest of the theorem follows from Theorem 5.1.1, Theorem 4.5.5(iii), and the definition of the integral (see Exercises 5-1, No. 1). □

Theorem 5.1.3
If $f : I \to \mathbb{R}$ is integrable over I with respect to α, then

$$|\int_I f\, d\alpha| \leq \int_I |f|\, d\alpha.$$

Proof By definition,

$$\left|\int_I f\, d\alpha\right| = |L_\alpha(f^+) - L_\alpha(-f^-)|;$$

thus,

$$\begin{aligned}\left|\int_I f\, d\alpha\right| &\leq L_\alpha(f^+) + L_\alpha(-f^-)\\ &= \int_I f^+\, d\alpha - \int_I f^-\, d\alpha = \int_I |f|\, d\alpha,\end{aligned}$$

by Theorem 5.1.1. □

Theorem 5.1.4 (Linearity of the Integral)
If for each $j = 1, 2, \ldots, m$, $f_j : I \to \mathbb{R}$ is integrable over I with respect to α, and a_j is a finite real number, then $\sum_{j=1}^m a_j f_j$ is integrable over I with

respect to α, and

$$\int_I \left(\sum_{j=1}^m a_j f_j\right) d\alpha = \sum_{j=1}^m a_j \int_I f_j\, d\alpha.$$

Proof For each $j = 1, 2, \dots, m$ we know that there exists a sequence $\theta_{j1}, \theta_{j2}, \dots$ of α-summable step functions such that $L_\alpha(|f_j - \theta_{jn}|) \to 0$ as $n \to \infty$. By Theorem 4.5.6 we have that for each $n = 1, 2, \dots$,

$$\begin{aligned}\int_I \left(\sum_{j=1}^m a_j\theta_{jn}\right) d\alpha &= A_\alpha\left(\sum_{j=1}^m a_j\theta_{jn}\right)\\ &= \sum_{j=1}^m a_j A_\alpha(\theta_{jn}) = \sum_{j=1}^m a_j \int_I \theta_{jn}\, d\alpha. \qquad (5.4)\end{aligned}$$

Now for each $n = 1, 2, \dots$

$$\begin{aligned}0 \le L_\alpha\left(\left|\sum_{j=1}^m a_j f_j - \sum_{j=1}^m a_j\theta_{jn}\right|\right) &= L_\alpha\left(\left|\sum_{j=1}^m a_j(f_j - \theta_{jn})\right|\right)\\ &\le \sum_{j=1}^m |a_j| L_\alpha(|f_j - \theta_{jn}|),\end{aligned}$$

by Theorems 4.5.4 and 4.5.2(ii). Since $L_\alpha(|f_j - \theta_{jn}|) \to 0$ for each $j = 1, 2, \dots, m$, it follows that

$$L_\alpha\left(\left|\sum_{j=1}^m a_j f_j - \sum_{j=1}^m a_j\theta_{jn}\right|\right) \to 0,$$

and so $\sum_{j=1}^m a_j f_j$ is integrable over I with respect to α by definition. It follows also, by Theorem 5.1.2, that $\int_I \theta_{jn}\, d\alpha \to \int_I f_j\, d\alpha$ as $n \to \infty$, for each $j = 1, 2, \dots, m$, and that $\int_I(\sum_{j=1}^m a_j\theta_{jn})\, d\alpha \to \int_I(\sum_{j=1}^m a_j f_j)\, d\alpha$. Letting n tend to ∞ in equation (5.4) gives the result. □

Theorem 5.1.5

Let $f, g : I \to \mathbb{R}$ be functions integrable over I with respect to α.

(i) *If $f \ge 0$ on I, then $\int_I f\, d\alpha \ge 0$.*

(ii) *If $g \le f$ on I, then $\int_I g\, d\alpha \le \int_I f\, d\alpha$.*

Proof Part (i) follows at once, since if $f \geq 0$ on I, then $\int_I f\,d\alpha = L_\alpha(f) \geq 0$. For part (ii) we need only observe that

$$\int_I f\,d\alpha = \int_I \{g + (f - g)\}\,d\alpha = \int_I g\,d\alpha + \int_I (f - g)\,d\alpha,$$

by Theorem 5.1.4. Part (i) implies that $\int_I (f - g)\,d\alpha \geq 0$, and therefore $\int_I f\,d\alpha \geq \int_I g\,d\alpha$. □

Exercises 5-1:

1. Complete the proof of Theorem 5.1.2.
2. For any functions $f, g : I \to \mathbb{R}$ we define the functions $\max\{f, g\} : I \to \mathbb{R}$ and $\min\{f, g\} : I \to \mathbb{R}$ by

$$(\max\{f, g\})(x) = \max\{f(x), g(x)\} \quad \text{for each } x \in I,$$
$$(\min\{f, g\})(x) = \min\{f(x), g(x)\} \quad \text{for each } x \in I.$$

   Prove that $\max\{f, g\} = f + (g - f)^+$ and $\min\{f, g\} = f + (g - f)^-$, and deduce that if f and g are both integrable over I with respect to α, then so are $\max\{f, g\}$ and $\min\{f, g\}$.
3. Prove the "first mean value theorem for integrals": If $f : I \to \mathbb{R}$ is integrable over I with respect to α, and if $\mu_\alpha(I)$ is finite, and if c_1 and c_2 are finite real numbers such that $c_1 \leq f \leq c_2$ on I, then

$$c_1 \mu_\alpha(I) \leq \int_I f\,d\alpha \leq c_2 \mu_\alpha(I).$$

4. Prove that if for each $j = 1, 2, \ldots, m$ the function $f_j : I \to \mathbb{R}$ is integrable over I with respect to α, then

$$\left| \int_I \left(\sum_{j=1}^m f_j \right) d\alpha \right| \leq \sum_{j=1}^m \left(\int_I |f_j|\,d\alpha \right).$$

5.2 Null Functions and Null Sets

If $f : I \to \mathbb{R}$ is such that $L_\alpha(|f|) = 0$, we call f a **null function** (with respect to α).

Theorem 5.2.1
If $f : I \to \mathbb{R}$ is a null function with respect to α, then f is integrable over I with respect to α and

$$\int_I f d\alpha = \int_I |f| \, d\alpha = 0.$$

Proof The sequence $\theta_1, \theta_2, \ldots$ defined by $\theta_n = 0$ for all $n = 1, 2, \ldots$ is a sequence of α-summable functions such that $L_\alpha(|f - \theta_n|) = L_\alpha(|f|) = 0$ for all $n = 1, 2, \ldots$, and so f is integrable. By Theorem 4.5.3(i) we have $L_\alpha(f^+) = L_\alpha(-f^-) = 0$, and therefore $\int_I f d\alpha = \int_I |f| \, d\alpha = 0$ by definition of these integrals. □

Corollary 5.2.2
A function $f : I \to \mathbb{R}$ is null with respect to α if and only if f is integrable over I with respect to α, and $\int_I |f| \, d\alpha = 0$.

Now let S be any subset of $\mathbb{R}$. We define the **characteristic function** of S to be the function $\chi_S(x) : \mathbb{R} \to \mathbb{R}$ defined by

$$\chi_S(x) = \begin{cases} 1, & \text{if } x \in S, \\ 0, & \text{if } x \notin S. \end{cases} \tag{5.5}$$

We say that S is a **null set** (with respect to α) if χ_S is a null function (with respect to α). Since χ_S is nonnegative, it follows at once from Corollary 5.2.2 that χ_S is null if and only if $\int_{\mathbb{R}} \chi_S \, d\alpha = 0$. It also follows from Theorem 4.5.3 and the definition of a null function that any subset of a null set is a null set.

If $f : I \to \mathbb{R}$ is a function, and P is some property of f that holds everywhere in I except possibly on some null subset of I, we say that P holds **almost everywhere** on I (abbreviated to a.e.). For example, if f_n $(n = 1, 2, \ldots)$ and f are functions defined on I, then "$f_n \to f$ a.e." means that $f_n(x) \to f(x)$ for all $x \in I$ except possibly for values of x belonging to some null subset of I.

Theorem 5.2.3
(i) *If $f : I \to \mathbb{R}$ is null, then $f = 0$ a.e.*
(ii) *If $f : I \to \mathbb{R}$ is such that $f = 0$ a.e., then f is null.*
(iii) *If $f, g : I \to \mathbb{R}$ are such that $f = g$ a.e., and if f is integrable over I, then g is also integrable over I, and $\int_I f \, d\alpha = \int_I g \, d\alpha$.*

Proof (i) Assume that $f : I \to \mathbb{R}$ is null, so that $L_\alpha(|f|) = 0$. We define the sequence of sets $A_1, A_2, \ldots$ as follows:

$$A_1 = \{x : x \in I, |f(x)| \geq 1\},$$
$$A_n = \left\{x : x \in I, \frac{1}{n} \leq |f(x)| < \frac{1}{n-1}\right\},$$

for $n = 2, 3, \ldots$. Clearly, $x \in A_n \Rightarrow 1 \leq n|f(x)|$ $(n = 1, 2, \ldots)$, and so for each $n = 1, 2, \ldots$ we have $0 \leq \chi_{A_n} \leq n|f|$ on I, and therefore

$$\begin{aligned} 0 \leq L_\alpha(\chi_{A_n}) &\leq L_\alpha(n|f|) && \text{by Theorem 4.5.3} \\ &= nL_\alpha(|f|) && \text{by Theorem 4.5.2(ii).} \end{aligned}$$

Since f is null, it follows that $L_\alpha(\chi_{A_n}) = 0$ for all $n = 1, 2, \ldots$.

Now let $S = \{x : x \in I, f(x) \neq 0\}$. For any $x \in S$ there exists a unique positive integer N such that $x \in A_N$, and therefore

$$\chi_{A_n}(x) = \begin{cases} 1, & \text{if } n = N, \\ 0, & \text{if } n \neq N. \end{cases}$$

Thus $\chi_S(x) = 1 = \sum_{n=1}^{\infty} \chi_{A_n}(x)$ for each $x \in S$. On the other hand, if $x \notin S$, then $x \notin A_n$ for each $n = 1, 2, \ldots$, and so $\chi_S(x) = 0 = \sum_{n=1}^{\infty} \chi_{A_n}(x)$. Therefore, $\chi_S = \sum_{n=1}^{\infty} \chi_{A_n}$, and so by Theorem 4.5.4 we have

$$0 \leq L_\alpha(\chi_S) \leq \sum_{n=1}^{\infty} L_\alpha(\chi_{A_n}) = \sum_{n=1}^{\infty} 0 = 0,$$

and so χ_S is null, which proves part (i).

(ii) Assume that $f = 0$ a.e. Let $S = \{x : x \in I, f(x) \neq 0\}$. Then S is a null set. For each $n = 1, 2, \ldots$ define $B_n = \{x : x \in I, n - 1 < |f(x)| \leq n\}$. Since $B_n \subseteq S$, it follows that B_n is null for each $n = 1, 2, \ldots$. Now define $f_n = n\chi_{B_n}$ for each $n = 1, 2, \ldots$. For any $x \in S$ there is a unique positive integer N such that $x \in B_N$, and therefore

$$f_n(x) = \begin{cases} N, & \text{if } n = N, \\ 0, & \text{if } n \neq N. \end{cases}$$

Hence $\sum_{n=1}^{\infty} f_n(x) = N \geq |f(x)|$ (since $x \in B_N$), and so we can say that $|f(x)| \leq \sum_{n=1}^{\infty} f_n(x)$ for all $x \in S$. On the other hand, if $x \in I - S$, then $x \notin B_n$ for each $n = 1, 2, \ldots$, and so $\sum_{n=1}^{\infty} f_n(x) = 0 = f(x)$. Thus $\sum_{n=1}^{\infty} f_n$ converges on I, and $|f| \leq \sum_{n=1}^{\infty} f_n$ on I. By

Theorems 4.5.2(ii), 4.5.3, and 4.5.4 we then have

$$\begin{aligned} 0 \leq L_\alpha(|f|) &\leq L_\alpha\left(\sum_{n=1}^{\infty} f_n\right) \\ &\leq \sum_{n=1}^{\infty} L_\alpha(f_n) \\ &= \sum_{n=1}^{\infty} L_\alpha(n\chi_{B_n}) = \sum_{n=1}^{\infty} nL_\alpha(\chi_{B_n}) \\ &= \sum_{n=1}^{\infty} 0 = 0, \end{aligned}$$

since B_n is null for each $n = 1, 2, \ldots$. Thus, $L_\alpha(|f|) = 0$, and f is null as required.

(iii) Since $f = g$ a.e., we have $g - f = 0$ a.e., so by part (ii), $g - f$ is null. By Theorem 5.2.1 it follows that $g - f$ is integrable and $\int_I (g - f)\, d\alpha = 0$. But $g = f + (g - f)$ on I, so by Theorem 5.1.4, g is integrable and

$$\int_I g\, d\alpha = \int_I f\, d\alpha + \int_I (g - f)\, d\alpha = \int_I f\, d\alpha,$$

as required. □

Part (iii) of the preceding theorem is particularly important. It shows that changing the values of a function on a null set does not affect the integral of the function. Two functions that are equal almost everywhere can be regarded as identical in the context of integration theory.

Exercises 5-2:

1. Let $\alpha^* : \mathbb{R} \to \mathbb{R}$ be defined by $\alpha^*(x) = x$ for all $x \in \mathbb{R}$. Prove that any finite or countably infinite subset of $\mathbb{R}$ is null with respect to α^*.

2. Give an example of a finite subset $S \subseteq \mathbb{R}$ and a monotone increasing function $\alpha : \mathbb{R} \to \mathbb{R}$ such that S is not null with respect to α.

3. Let $g : [0, 1] \to \mathbb{R}$ be defined by

$$g(x) = \begin{cases} 0, & \text{if } x \text{ is rational, } x \neq 0, 1, \\ 1, & \text{if } x \text{ is irrational or } x = 0 \text{ or } x = 1. \end{cases}$$

   Find $\int_{[0,1]} g \, dx$, explaining your reasoning in full.

4. Generalize Theorem 5.1.5(ii) by proving that if $f, g : I \to \mathbb{R}$ are integrable over I with respect to α, and $g \leq f$ a.e., then $\int_I g \, d\alpha \leq \int_I f \, d\alpha$.

5. Prove that the union of two null sets is a null set.

5.3 Convergence Theorems

We now state the main convergence theorems for the Lebesgue–Stieltjes integral. The proofs are, regrettably, too long and technically difficult to include here. The reader is referred to [31], [32], or [38] for the details.

Theorem 5.3.1 (Monotone Convergence Theorem)
Let $f_1, f_2, \ldots$ be a monotone sequence of functions that are all integrable over I with respect to α, and are such that $\lim_{n\to\infty}(\int_I f_n \, d\alpha)$ is finite. Let $f : I \to \mathbb{R}$ be such that $f_n \to f$ a.e. Then f is also integrable over I with respect to α, and $\int_I f_n \, d\alpha \to \int_I f \, d\alpha$.

Suppose that $\{a_n\}$ is a sequence of real numbers bounded below. Let $k_n = \inf_{m\geq n} a_m$. Then by Exercises 1-3, No. 3, the sequence $\{k_n\}$ is monotone increasing, and so by Theorem 2.1.1,

$$\lim_{n\to\infty} k_n = \lim_{n\to\infty} (\inf_{m\geq n} a_m)$$

exists in $\mathbb{R}_e$. A consequence of the monotone convergence theorem is the following technical result, which we shall use in Chapter 8:

Lemma 5.3.2 (Fatou's Lemma)
Let $f_1, f_2, \ldots$ be a sequence of nonnegative functions that are all integrable over the interval I with respect to α and suppose that $f_n(x) \to f(x)$ for all $x \in I$ except perhaps in a null set of I. Then f is integrable over I with respect to α if and only if $\lim_{n\to\infty} \left(\inf_{m\geq n} \int_I f_m(x) \, d\alpha\right)$ is finite, and in

that case

$$\int_I f\,d\alpha \le \lim_{n\to\infty}\left(\inf_{m\ge n}\int_I f_m(x)\,d\alpha\right).$$

Theorem 5.3.3 (Dominated Convergence Theorem)
Let $f_1, f_2, \ldots$ be a monotone sequence of functions that are all integrable over I with respect to α and are such that for each $n = 1, 2, \ldots$, $|f_n| \le \lambda$ on I, where $\lambda : I \to \mathbb{R}$ is integrable over I with respect to α. Let $f : I \to \mathbb{R}$ be such that $f_n \to f$ a.e. Then f is also integrable over I with respect to α, and we have

$$\int_I f_n d\alpha \to \int_I f\,d\alpha \quad \text{and} \quad \int_I |f_n - f|\,d\alpha \to 0.$$

As immediate consequences of Theorems 5.3.1 and 5.3.3, obtained by applying these theorems to the sequence of partial sums of a series (see Exercises 5-3, No. 1), we obtain the following important results on the integration of series term by term.

Theorem 5.3.4
Let $a_1, a_2, \ldots$ be a sequence of functions that are all integrable over I with respect to α, all have the same sign (i.e., either $a_j \ge 0$ for all $j = 1, 2, \ldots$ or $a_j \le 0$ for all $j = 1, 2, \ldots$), and are such that $\sum_{j=1}^{\infty}(\int_I a_j\,d\alpha)$ converges. Let $s : I \to \mathbb{R}$ be such that $\sum_{j=1}^{\infty} a_j = s$ a.e. Then s is also integrable over I with respect to α, and

$$\int_I s\,d\alpha = \sum_{j=1}^{\infty}(\int_I a_j\,d\alpha).$$

Theorem 5.3.5
Let $a_1, a_2, \ldots$ be a sequence of functions that are all integrable over I with respect to α, and are such that for each $n = 1, 2, \ldots,$

$$\left|\sum_{j=1}^{n} a_j\right| \le \lambda$$

on I, where $\lambda : I \to \mathbb{R}$ is integrable over I with respect to α. Let $s : I \to \mathbb{R}$ be such that $\sum_{j=1}^{\infty} a_j = s$ a.e. Then s is also integrable over I with respect

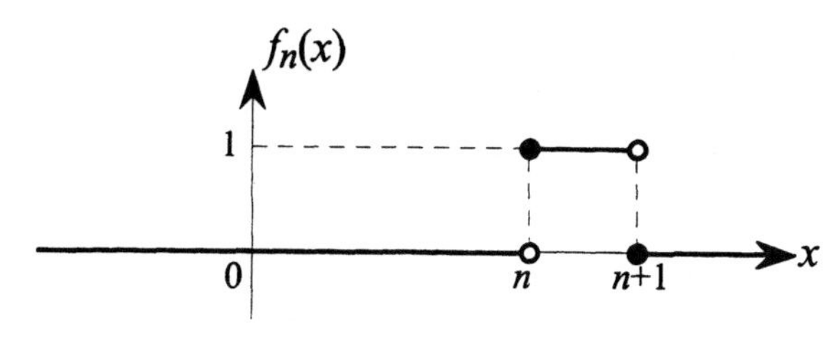

FIGURE 5.1

to α, and

$$\int_I s\,d\alpha = \sum_{j=1}^{\infty}\left(\int_I a_j\,d\alpha\right).$$

Exercises 5-3:

1. (a) Deduce Theorem 5.3.4 from Theorem 5.3.1.
   (b) Deduce Theorem 5.3.5 from Theorem 5.3.3.
2. For each $n = 1, 2, \ldots$ let $f_n : \mathbb{R} \to \mathbb{R}$ be defined by

$$f_n(x) = \begin{cases} 1, & \text{if } n \le x < n+1, \\ 0, & \text{otherwise} \end{cases}$$

   (Figure 5.1).
   (a) Find the function $f = \lim_{n\to\infty} f_n$.
   (b) Show that $\lim_{n\to\infty}(\int_{\mathbb{R}} f_n\,dx) \neq \int_{\mathbb{R}} f\,dx$.
   (c) State which hypothesis of the monotone convergence theorem is not satisfied in this case, and explain in detail why it is not satisfied. Do the same for the dominated convergence theorem.

5.4 Extensions of the Theory

In this section we describe briefly two important ways in which the definition of $\int_I f d\alpha$ can be extended: (1) by allowing integration over sets other than intervals, and (2) by allowing α to be a function of bounded variation.

(1) We say that a **function** $f : I \to \mathbb{R}$ is α-**measurable** if there is a sequence $\theta_1, \theta_2, \ldots$ of α-summable step functions on I such that $\lim_{n\to\infty} \theta_n = f$ a.e. Any α-summable step function is obviously α-measurable, and so is the unit function 1, which takes the value 1 for each $x \in \mathbb{R}$ (see Exercises 5-4, No. 1). From the properties of step functions and the elementary rules for limits, it follows easily that:

(i) the modulus and the positive and negative parts of an α-measurable function are α-measurable;

(ii) if $f, g : I \to \mathbb{R}$ are α-measurable, then so are $f \pm g$, fg, $\max\{f, g\}$, and $\min\{f, g\}$. It is also true that f/g is α-measurable, provided that $g \neq 0$ a.e., but this is a little harder to prove.

It can be proved that a function $f : I \to \mathbb{R}$ is integrable over I with respect to α if and only if it is α-measurable and $L_\alpha(|f|)$ is finite.

A **set** $X \subseteq \mathbb{R}$ is said to be α-**measurable** if its characteristic function χ_X is α-measurable. If X is α-measurable and $L_\alpha(\chi_X)$ is finite, then χ_X is integrable over $\mathbb{R}$ with respect to α, and we define the α-**measure** $\mu_\alpha(X)$ of X by

$$\mu_\alpha(X) = \int_{\mathbb{R}} \chi_X \, d\alpha.$$

If X is α-measurable and $L_\alpha(\chi_X) = \infty$, we say that X has **infinite** α-**measure** and write $\mu_\alpha(X) = \infty$.

This definition of measure is easily seen to be consistent with our previous definition of measure for simple sets. Note also that a set is null with respect to α if and only if it has α-measure zero. The concept of the measure of a set is of considerable importance in itself, quite apart from its connection with integration.

If a function $f : \mathbb{R} \to \mathbb{R}$ is integrable over $\mathbb{R}$ with respect to α, and $X \subseteq \mathbb{R}$ is α-measurable, then $f\chi_X$ is α-measurable, and since $|f\chi_X| \leq |f|$ on $\mathbb{R}$, we can say that $L_\alpha(|f\chi_X|)$ is finite and so $f\chi_X$ is integrable over $\mathbb{R}$ with respect to α. We can then define the **integral of** f **over** X **with respect to** α by

$$\int_X f \, d\alpha = \int_{\mathbb{R}} f\chi_X \, d\alpha.$$

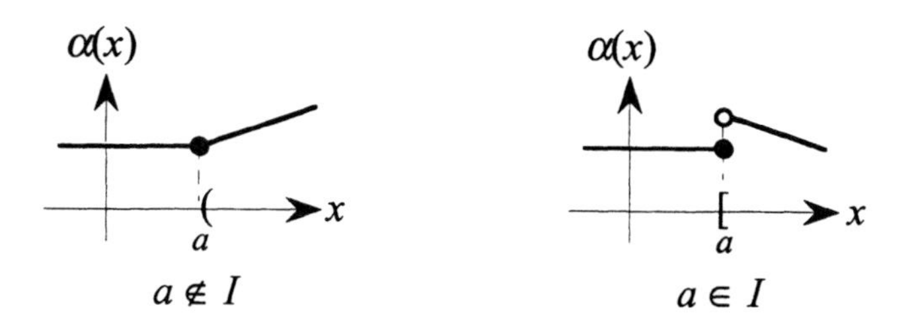

FIGURE 5.2

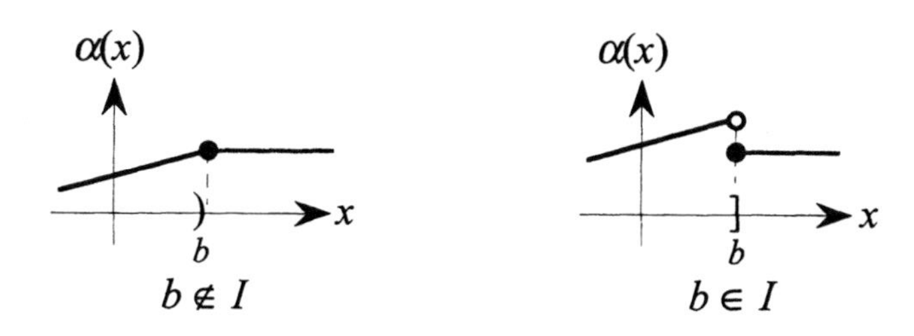

FIGURE 5.3

(2) Suppose $\alpha : I \to \mathbb{R}$ is a function of bounded variation. If I is a proper subset of $\mathbb{R}$ with endpoints a, b, then we have by Corollary 2.7.3 that $\alpha(a^+)$ and $\alpha(b^-)$ exist and are finite. We extend α to a function of bounded variation on $\mathbb{R}$ in the following way: (i) If a is finite, then define $\alpha(x)$ to be equal to $\alpha(a^+)$ for all $x \leq a$ in the case where $a \notin I$, and equal to $\alpha(a)$ for all $x < a$ in the case where $a \in I$ (Figure 5.2). (ii) If b is finite, then define $\alpha(x)$ to be equal to $\alpha(b^-)$ for all $x \geq b$ in the case where $b \notin I$, and equal to $\alpha(b)$ for all $x > b$ in the case $b \in I$ (Figure 5.3). By Theorem 2.7.4, we can then express α as a difference

$$\alpha = \alpha_1 - \alpha_2,$$

where $\alpha_1, \alpha_2 : \mathbb{R} \to \mathbb{R}$ are both monotone increasing.

Now let J be any subinterval of I. If a function $f : J \to \mathbb{R}$ is integrable over J with respect to both α_1 and α_2, we say that f is integrable over J with respect to α, and make the natural definition

$$\int_J f\,d\alpha = \int_J f\,d\alpha_1 - \int_J f\,d\alpha_2.$$

It can be proved that the value of $\int_J f\,d\alpha$ does not depend on the particular way in which α is expressed as a difference of monotone increasing functions.

In the same spirit, we define the α-measure of an interval $J \subseteq I$ to be

$$\mu_\alpha(J) = \mu_{\alpha_1}(J) - \mu_{\alpha_2}(J).$$

From the elementary rules for limits, it is easy to see that $\mu_\alpha(J)$ can be described in terms of the one-sided limits of α at the endpoints of J in precisely the same way as was done for monotone increasing α (Section 4-1). Of course, when α is a function of bounded variation, we must allow for the possibility that intervals may have negative α-measure. Note also (and particularly) that the theory of null sets and null functions, as developed in Section 5-2, is no longer valid in this more general setting.

For the most part we will continue to restrict ourselves to integration with respect to monotone increasing functions, but we will need the extension to functions of bounded variation when we discuss integration by parts in the next chapter.

Exercises 5-4:

1. Prove that the unit function $1 : \mathbb{R} \to \mathbb{R}$ is α-measurable for any monotone increasing function $\alpha : \mathbb{R} \to \mathbb{R}$.
2. Define the functions $f, \alpha : [0, 2) \to \mathbb{R}$ as follows:

$$f(x) = \begin{cases} 1, & \text{if } 0 \leq x < 1, \\ -1, & \text{if } 1 \leq x < 2, \end{cases}$$

$$\alpha(x) = \begin{cases} x, & \text{if } 0 \leq x \leq 1, \\ 1 - x, & \text{if } 1 < x < 2 \end{cases}$$

(Figure 5.4).

(a) Extend α to a function on $\mathbb{R}$ as specified in Section 5-4. Show that α, so extended, is a function of bounded variation on $\mathbb{R}$,

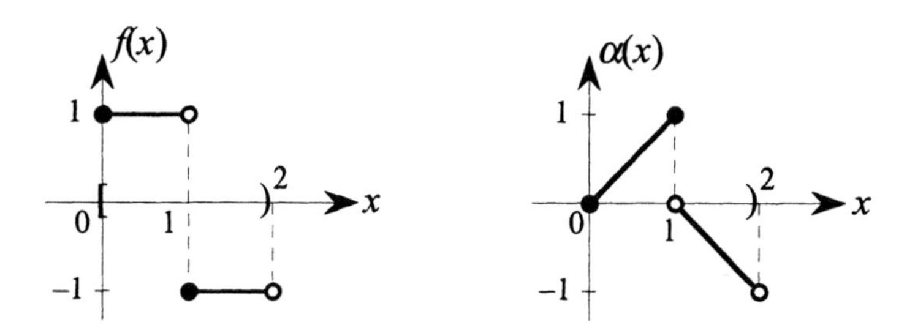

FIGURE 5.4

by expressing α as a difference of two monotone increasing functions $\alpha_1, \alpha_2 : \mathbb{R} \to \mathbb{R}$ that satisfy the hypotheses of Theorem 2.7.2.

(b) Find $\mu_\alpha([0, 2))$, $\mu_\alpha([0, 1])$, $\mu_\alpha([1, 1])$ and $\mu_\alpha((1, 2))$.

(c) Evaluate $\int_{[0,2)} f \, d\alpha$.

6 CHAPTER Integral Calculus

Having worked through the basic theory of the integral, we now turn to the actual techniques of integral calculus, that is, the practicalities of evaluating and manipulating Lebesgue–Stieltjes integrals. The emphasis here will be on understanding and applying the results. For this reason, the proofs of most the results stated in this chapter will be suppressed. The results are standard to the theory, and their proofs can be found in most texts on integration (e.g., [31], [32], [38]). It should be noted, however, that many of these proofs pose a technical demand on the reader greater than what has been expected so far in this text.

6.1 Evaluation of Integrals

The actual evaluation of Lebesgue–Stieltjes integrals takes as its starting point the fact that for a closed interval $[a, b]$ the ordinary Riemann integral $\int_a^b f(x)\,dx$ (if it exists) has the same value as the Lebesgue integral $\int_{[a,b]} f\,dx$ (cf. Theorem 4.6.1). The same holds for improper Riemann integrals, provided that they are absolutely convergent (as discussed at the end of Chapter 4). Lebesgue integrals that correspond to Riemann integrals can therefore be evaluated using all the elementary techniques with which you are presumed familiar. More general Lebesgue–Stieltjes integrals can usually be dealt with by reducing them to combinations of integrals that either are equivalent to Riemann integrals or are easy to evaluate. The theorems that follow provide the necessary tools. In all of them, we assume that $\alpha : \mathbb{R} \to \mathbb{R}$ is a monotone increasing function.

Theorem 6.1.1
If the interval I is a union of a finite number of pairwise disjoint intervals

$$I = I_1 \cup I_2 \cup \cdots \cup I_n,$$

then

$$\int_I f\, d\alpha = \sum_{j=1}^{n} \int_{I_j} f\, d\alpha$$

in the sense that if one side exists, then so does the other, and the two are equal.

Theorem 6.1.2
Let $\alpha = \sum_{j=1}^{m} c_j \alpha_j$, where for each $j = 1, 2, \ldots, \alpha_j : \mathbb{R} \to \mathbb{R}$ is a monotone increasing function and c_j is a nonnegative finite real number. If a function $f : I \to \mathbb{R}$ is integrable over I with respect to each of $\alpha_1, \alpha_2, \ldots, \alpha_m$, then it is integrable over I with respect to α, and

$$\int_I f\, d\alpha = \sum_{j=1}^{m} c_j \int_I f\, d\alpha_j.$$

Theorem 6.1.3

(i) *If α is continuous at a, then*

$$\int_{[a,b]} f\, d\alpha = \int_{(a,b]} f\, d\alpha \quad \text{and} \quad \int_{[a,b)} f\, d\alpha = \int_{(a,b)} f\, d\alpha$$

in the sense that if one side of the equation exists, then so does the other, and the two are equal.

(ii) *If α is continuous at b, then*

$$\int_{[a,b]} f\, d\alpha = \int_{[a,b)} f\, d\alpha \quad \text{and} \quad \int_{(a,b]} f\, d\alpha = \int_{(a,b)} f\, d\alpha$$

in the sense that if one side of the equation exists, then so does the other, and the two are equal.

Theorem 6.1.4
For any interval I, $\int_I 1\, d\alpha = \mu_\alpha(I)$.

Theorem 6.1.5
If α *is constant on an open interval I, then* $\int_I f\, d\alpha = 0$.

Theorem 6.1.6
For any function f defined at a, $\int_{[a,a]} f\, d\alpha = f(a)[\alpha(a^+) - \alpha(a^-)]$.

Theorem 6.1.7
If α *is differentiable at all points in an open interval I, then*

$$\int_I f\, d\alpha = \int_I f\alpha'\, dx$$

in the sense that if one side exists, then so does the other, and the two are equal.

Theorem 6.1.8
Let I be an open interval, and let $\beta : I \to \mathbb{R}$ *be a monotone increasing function on I such that* $\alpha(x) = \beta(x)$ *for all* $x \in I$. *Then*

$$\int_I f\, d\alpha = \int_I f\, d\beta$$

(note that since I is open, the integral on the right is defined even if the domain of β *does not extend beyond I).*

A few examples should be sufficient to show how these theorems are applied. The most important thing to note is that points of discontinuity of α must be dealt with separately, by treating them as single-point closed intervals and using Theorem 6.1.6.

Example 6-1-1:
Let $\alpha : \mathbb{R} \to \mathbb{R}$ be defined by

$$\alpha(x) = \begin{cases} 0, & \text{if } x < 2, \\ 1, & \text{if } x \geq 2 \end{cases}$$

(cf. Figure 6.1).

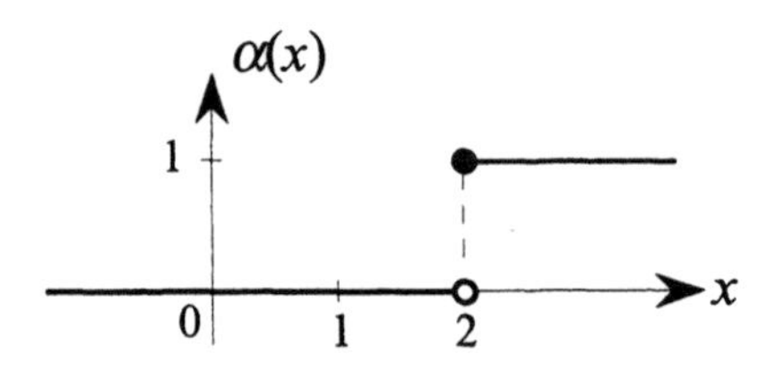

FIGURE 6.1

Then $\int_{[2,3]} x^2\, d\alpha = \int_{[2,2]} x^2\, d\alpha + \int_{(2,3]} x^2\, d\alpha$ by Theorem 6.1.1 (note that α is discontinuous at 2). Now, Theorem 6.1.6 implies that

$$\int_{[2,2]} x^2\, d\alpha = 2^2(\alpha(2^+) - \alpha(2^-)) = 4(1-0) = 4,$$

and Theorem 6.1.3(ii) implies that

$$\int_{(2,3]} x^2\, d\alpha = \int_{(2,3)} x^2\, d\alpha.$$

The above equation and Theorem 6.1.5 indicate that

$$\int_{(2,3]} x^2\, d\alpha = 0,$$

and thus

$$\int_{[2,3]} x^2\, d\alpha = 4 + 0 = 4.$$

Note that Theorem 6.1.5 requires the interval to be open, and we used Theorem 6.1.3(ii) to change the interval of integration to the open interval $(2,3)$. The main feature to note from this example is that although α is constant on the *closed* interval $[2,3]$, we cannot conclude that $\int_{[2,3]} x^2\, d\alpha = 0$, because of the discontinuity in α.

Example 6-1-2:
Let $\alpha : \mathbb{R} \to \mathbb{R}$ be defined by

$$\alpha(x) = \begin{cases} 0, & \text{if } x < 0, \\ 3 - e^{-2x}, & \text{if } x \geq 0 \end{cases}$$

(Figure 6.2).

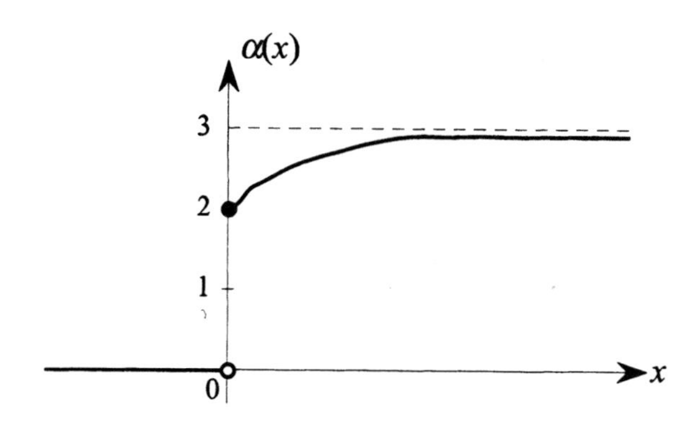

FIGURE 6.2

Then,

$$\int_{[0,\infty)} e^x \, d\alpha = \int_{[0,0]} e^x \, d\alpha + \int_{(0,\infty)} e^x \, d\alpha \quad \text{(by Theorem 6.1.1)}$$

$$= e^0(\alpha(0^+) - \alpha(0^-)) + \int_{(0,\infty)} e^x \, d(3 - e^{-2x})$$

(by Theorems 6.1.6 and 6.1.8).

Applying Theorem 6.1.7 to the second integral gives

$$\begin{aligned}
\int_{[0,\infty)} e^x \, d\alpha &= 1(2-0) + \int_{(0,\infty)} e^x(2e^{-2x}) \, dx \\
&= 2 + \int_{[0,\infty)} 2e^{-x} \, dx \quad \text{(by Theorem 6.1.3(i))} \\
&= 2 + \int_0^\infty 2e^{-x} \, dx \quad \text{(an improper Riemann integral)} \\
&= 2 + \lim_{c\to\infty} \left[-2e^{-x} \right]_0^c \\
&= 2 + (0 - (-2)) = 4.
\end{aligned}$$

Note again that although $\alpha(x) = 3 - e^{-2x}$ for all $x \in [0,\infty)$, we cannot conclude that $\int_{[0,\infty)} e^x \, d\alpha = \int_{[0,\infty)} e^x \, d(3 - e^{-2x})$. Again, the discontinuity in α at the endpoint of the interval is crucial; the condition in Theorem 6.1.8 that I must be *open* cannot be disregarded.

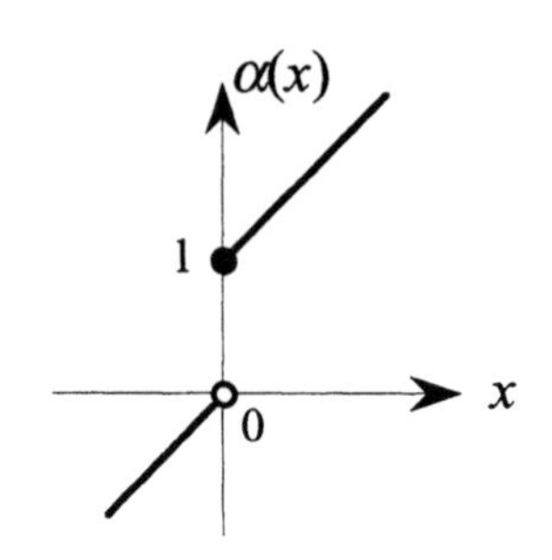

FIGURE 6.3

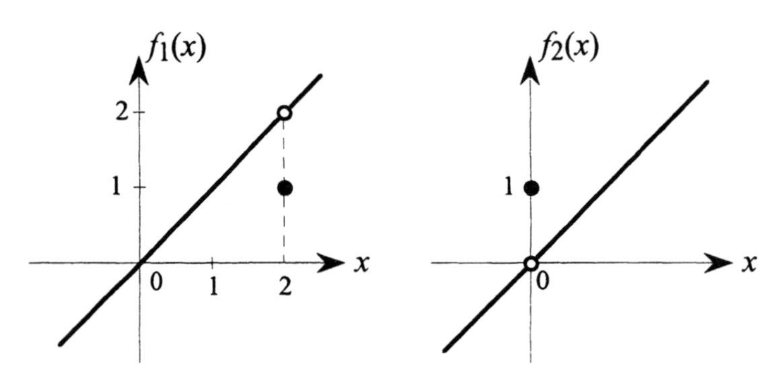

FIGURE 6.4

Example 6-1-3:
Let $\alpha : \mathbb{R} \to \mathbb{R}$ be defined by

$$\alpha(x) = \begin{cases} x, & \text{if } x < 0, \\ x+1, & \text{if } x \geq 0 \end{cases}$$

(Figure 6.3).

Let $f_1, f_2 : \mathbb{R} \to \mathbb{R}$ be defined by

$$f_1(x) = \begin{cases} x, & \text{if } x \neq 2, \\ 1, & \text{if } x = 2, \end{cases} \qquad f_2(x) = \begin{cases} x, & \text{if } x \neq 0, \\ 1, & \text{if } x = 0 \end{cases}$$

(Figure 6.4). Now, the set $\{2\}$ has zero α-measure, and Theorem 5.2.3 implies that $\int_{[1,3]} f_1 \, d\alpha = \int_{[1,3]} x \, d\alpha$. Thus,

$$\int_{[1,3]} f_1 \, d\alpha = \int_{(1,3)} x \, d(x+1) \quad \text{(by Theorems 6.1.3 and 6.1.8)}$$

$$= \int_{(1,3)} x(1)\,dx \quad \text{(by Theorem 6.1.7)}$$

$$= \int_1^3 x\,dx \quad \text{(a Riemann integral) by Theorem 6.1.3}$$

$$= \left[\frac{x^2}{2}\right]_1^3 = \frac{9}{2} - \frac{1}{2} = 4.$$

However, we cannot say that $\int_{[-1,1]} f_2\,d\alpha = \int_{[-1,1]} x\,d\alpha$, because the set $\{0\}$ does not have zero α-measure. Instead, we proceed thus:

$$\int_{[-1,1]} f_2\,d\alpha = \int_{[-1,0)} f_2\,d\alpha + \int_{[0,0]} f_2\,d\alpha + \int_{(0,1]} f_2\,d\alpha$$

(by Theorem 6.1.1)

$$= \int_{-1}^0 x\,dx + f_2(0)(\alpha(0^+) - \alpha(0^-)) + \int_0^1 x\,dx$$

(by Theorem 6.1.6)

(using reasoning similar to that used in the first part to deal with the first and third integrals). Therefore,

$$\int_{[-1,1]} f_2\,d\alpha = \left[\frac{x^2}{2}\right]_{-1}^0 + 1(1-0) + \left[\frac{x^2}{2}\right]_0^1 = 1.$$

Note that in contrast, $\int_{[-1,1]} x\,d\alpha = 0$, the calculation being the same as the preceding one except that $f_2(0)$ is replaced by 0, the value of the integrand at 0.

Example 6-1-4:
Let $\alpha : \mathbb{R} \to \mathbb{R}$ be defined by

$$\alpha(x) = \begin{cases} 0, & \text{if } x < 1, \\ x^2 - 2x + 2, & \text{if } 1 \le x < 2, \\ 3, & \text{if } x = 2, \\ x + 2, & \text{if } x > 2 \end{cases}$$

(Figure 6.5). Then,

$$\int_{[0,3)} x^2\,d\alpha = \int_{[0,1)} x^2\,d\alpha + \int_{[1,1]} x^2\,d\alpha + \int_{(1,2)} x^2\,d\alpha$$

$$+ \int_{[2,2]} x^2\,d\alpha + \int_{(2,3)} x^2\,d\alpha \quad \text{(by Theorem 6.1.1)}$$

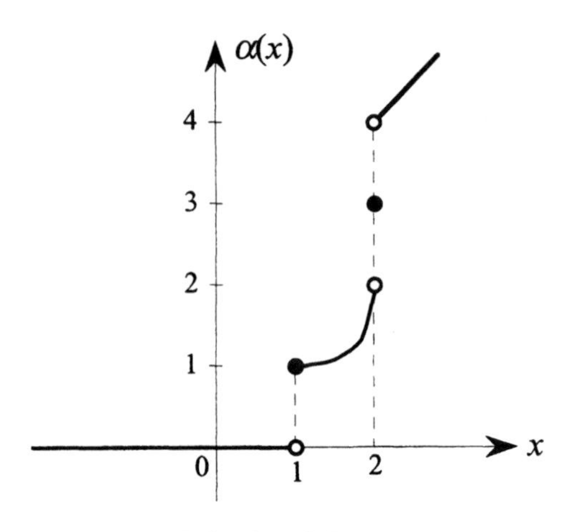

FIGURE 6.5

$$= \int_{(0,1)} x^2\, d\alpha + 1^2(\alpha(1^+) - \alpha(1^-)) + \int_{(1,2)} x^2\, d(x^2 - 2x + 2)$$

$$+ 2^2(\alpha(2^+) - \alpha(2^-)) + \int_{(2,3)} x^2\, d(x + 2)$$

(by Theorems 6.1.3 (i), 6.1.6, and 6.1.8)

$$= 0 + 1(1 - 0) + \int_{(1,2)} x^2(2x - 2)\, dx + 4(4 - 2)$$

$$+ \int_{(2,3)} x^2 1\, dx \quad \text{(by Theorems 6.1.5 and 6.1.7)}$$

$$= 1 + \int_{[1,2]} (2x^3 - 2x^2)\, dx + 8 + \int_{[2,3]} x^2\, dx$$

(by Theorem 6.1.3)

$$= 9 + \int_1^2 (2x^3 - 2x^2)\, dx + \int_2^3 x^2\, dx \quad \text{(Riemann integrals)}$$

$$= 9 + \left[\frac{x^4}{2} - \frac{2x^3}{3}\right]_1^2 + \left[\frac{x^3}{3}\right]_2^3$$

$$= \frac{109}{6}.$$

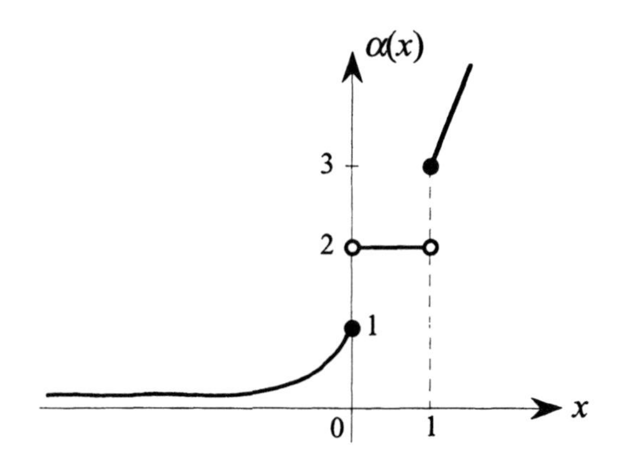

FIGURE 6.6

Exercises 6-1:

1. Let $\alpha : \mathbb{R} \to \mathbb{R}$ be defined by

$$\alpha(x) = \begin{cases} e^{3x}, & \text{if } x \leq 0, \\ 2, & \text{if } 0 < x < 1, \\ 2x + 1, & \text{if } x \geq 1 \end{cases}$$

(cf. Figure 6.6).

Let $f : \mathbb{R} \to \mathbb{R}$ be defined by

$$f(x) = \begin{cases} e^{-2x}, & \text{if } x \leq 1, \\ x, & \text{if } x > 1 \end{cases}$$

(cf. Figure 6.7). Evaluate $\int_I f \, d\alpha$ for each of the following intervals I:

(i) $(-1, 0)$ (ii) $[-1, 0]$ (iii) $(-1, 1)$
(iv) $(-1, 1]$ (v) $[1, 3]$ (vi) $(-\infty, 0)$

2. Let $[x]$ denote the integer part of x, i.e., the largest integer n such that $n \leq x$; for example, $[2.71] = 2$, $[3] = 3$, $[-1.82] = -2$.

   (a) Sketch the graph of the function $[x] : \mathbb{R} \to \mathbb{R}$.

   (b) Evaluate the following integrals:

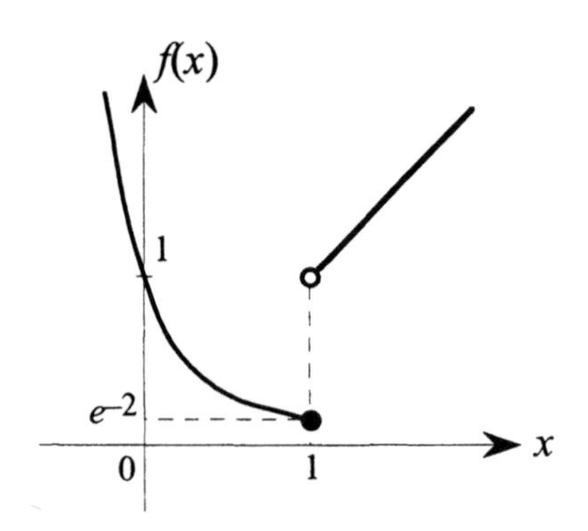

FIGURE 6.7

(i) $\int_{[0,5]}(x^2+1)\,d[x]$ (ii) $\int_{[0,5]} e^x\,d(x+[x])$

(iii) $\int_{[1/4,5/4]}[x]\,d[2x]$ (iv) $\int_{[1/4,5/4]}[2x]\,d[x]$

3. Let $\alpha : \mathbb{R} \to \mathbb{R}$ be a probability distribution function corresponding to a random variable x. We define the *mean* of x (also called the *expectation* or the *expected value*) to be

$$E(x) = \int_{(-\infty,\infty)} x\,d\alpha.$$

   (a) Calculate the mean of the uniformly distributed random variable defined in Example 4-2-1.

   (b) Calculate the means of the random variables defined in Example 4-2-2 and Exercises 4-2, No. 2.

   (c) If x is a random variable that can take exactly n values $\lambda_1, \lambda_2, \ldots, \lambda_n$ (where $\lambda_1 < \lambda_2 < \cdots < \lambda_n$) with probabilities $p_1, p_2, \ldots, p_n$, respectively (where $\sum_{j=1}^{n} p_j = 1$), find the corresponding probability distribution function α, and the mean of this random variable.

6.2 Two Theorems of Integral Calculus

In this section we will look at the form taken in the Lebesgue–Stieltjes theory by two theorems that are familiar to you in the context of elementary integral calculus.

An important aid in evaluating elementary integrals is the "change of variable" theorem

$$\int_a^b f[u(t)]u'(t)\,dt = \int_{u(a)}^{u(b)} f(x)\,dx,$$

where we have made the substitution $x = u(t)$ in order to simplify the integral on the left. The same thing can be done within the Lebesgue–Stieltjes theory.

First we need a definition. We say that a function $u : \mathbb{R} \to \mathbb{R}$ is **strictly increasing** on an interval I if $u(x_1) < u(x_2)$ for all $x_1, x_2 \in I$ such that $x_1 < x_2$. Suppose now that $u : \mathbb{R} \to \mathbb{R}$ is continuous and strictly increasing on I, and write

$$u(I) = \{u(x) : x \in I\}.$$

Then it is easy to see that $u(I)$ is an interval. For example, if $u(x) = x^3 + 1$ for all $x \in \mathbb{R}$, then $u((-2, 1)) = (-7, 2)$, $u([3, \infty)) = [28, \infty)$, $u((-\infty, -1)) = (-\infty, 0)$, and so on.

We can now state the "change of variable" theorem for the Lebesgue–Stieltjes integral.

Theorem 6.2.1 (Change of Variable)
Let I be any interval, and $u : \mathbb{R} \to \mathbb{R}$ be a function that is continuous and strictly increasing on the interval I. Then

$$\int_I (f \circ u)\,du = \int_{u(I)} f\,dx,$$

where $f \circ u$ denotes the composition of f and u, defined by $(f \circ u)(x) = f[u(x)]$ for all $x \in I$. If, in addition, u is differentiable on I, then this result can be written in the form

$$\int_I (f \circ u)u'\,dx = \int_{u(I)} f\,dx.$$

Finally, if $\alpha : \mathbb{R} \to \mathbb{R}$ *is monotone increasing, then*

$$\int_I (f \circ u)\, d(\alpha \circ u) = \int_{u(I)} f\, d\alpha.$$

All three results hold in the sense that if one side exists, then so does the other, and the two are equal.

The condition that u should be strictly increasing on I is not really a restriction in practice. If u is not strictly increasing, the interval of integration can usually be split up into subintervals on which u is either strictly increasing or strictly decreasing, or constant, and each of these can be dealt with separately (note that if u is strictly decreasing then $-u$ is strictly increasing, so the theorem can still be used with the obvious modifications). When evaluating integrals in practice, therefore, one is usually safe, provided that u is continuous (and for the most part u is also differentiable).

The second matter we will consider here is integration by parts. Recall that for the Riemann integral, the technique of integration by parts centers on the formula

$$\int_a^b fg'\, dx + \int_a^b f'g\, dx = [fg]_a^b.$$

For the Lebesgue–Stieltjes integral, this result takes the basic form

$$\int_I f\, dg + \int_I g\, df = \mu_{fg}(I),$$

where f and g can be allowed to be functions of bounded variation, using the approach outlined in Chapter 5. However, a correction term is needed in order to take account of cases where f and g have points of discontinuity in common.

In order to understand the need for a correction term, it is sufficient to consider the simplest case, where I is the closed interval $[a, a] = \{a\}$ consisting of a single point. We then have

$$\int_{\{a\}} f\, dg = f(a)[g(a^+) - g(a^-)] = f(a)\mu_g(\{a\}),$$

$$\int_{\{a\}} g\, df = g(a)[f(a^+) - f(a^-)] = g(a)\mu_f(\{a\}),$$

$$\mu_{fg}(\{a\}) = (fg)(a^+) - (fg)(a^-) = f(a^+)g(a^+) - f(a^-)g(a^-).$$

Thus,

$$\begin{aligned}
&\left(\int_{\{a\}} f\,dg + \int_{\{a\}} g\,df\right) - \mu_{fg}(\{a\}) \\
&\quad = f(a)\mu_g(\{a\}) + g(a)\mu_f(\{a\}) - \left[f(a^+)g(a^+) - f(a^-)g(a^-)\right] \\
&\quad = f(a)\mu_g(\{a\}) + g(a)\mu_f(\{a\}) \\
&\qquad - \left[f(a^+)g(a^+) - f(a^+)g(a^-) + f(a^+)g(a^-) - f(a^-)g(a^-)\right] \\
&\quad = f(a)\mu_g(\{a\}) + g(a)\mu_f(\{a\}) - \left[f(a^+)\mu_g(\{a\}) - g(a^-)\mu_f(\{a\})\right] \\
&\quad = (f(a) - f(a^+))\mu_g(\{a\}) + (g(a) - g(a^-))\mu_f(\{a\}) = A(a), \text{ say.}
\end{aligned}$$

Then also

$$\begin{aligned}
A(a) &= \left[f(a) - f(a^-) + f(a^-) - f(a^+)\right]\mu_g(\{a\}) \\
&\quad + \left[g(a) - g(a^+) + g(a^+) - g(a^-)\right]\mu_f(\{a\}) \\
&= \left[f(a) - f(a^-)\right]\mu_g(\{a\}) - \mu_f(\{a\})\mu_g(\{a\}) \\
&\quad + \left[g(a) - g(a^+)\right]\mu_f(\{a\}) + \mu_g(\{a\})\mu_f(\{a\}) \\
&= \left[f(a) - f(a^-)\right]\mu_g(\{a\}) + \left[g(a) - g(a^+)\right]\mu_f(\{a\}).
\end{aligned}$$

Adding gives

$$2A(a) = \left[2f(a) - f(a^+) - f(a^-)\right]\mu_g(\{a\}) + \left[2g(a) - g(a^-) - g(a^+)\right]\mu_f(\{a\}),$$

and so finally,

$$\begin{aligned}
A(a) &= \left[f(a) - \frac{1}{2}(f(a^+) + f(a^-))\right]\mu_g(\{a\}) \\
&\quad + \left[g(a) - \frac{1}{2}(g(a^+) + g(a^-))\right]\mu_f(\{a\}).
\end{aligned}$$

We then have

$$\int_{\{a\}} f\,dg + \int_{\{a\}} g\,df = \mu_{fg}(\{a\}) + A(a).$$

Note first that if f is continuous at a, then $f(a) = \frac{1}{2}(f(a^+) + f(a^-))$ and $\mu_f(\{a\}) = 0$, and so $A(a) = 0$; similarly if g is continuous at a. Thus it is possible for A to be nonzero only if both f and g are discontinuous at a. Three important special cases are:

(i) If $f(a) = \frac{1}{2}(f(a^+) + f(a^-))$ and $g(a) = \frac{1}{2}(g(a^+) + g(a^-))$, then $A(a) = 0$.

(ii) If f and g are both continuous on the right at a, so that $f(a^+) = f(a)$ and $g(a^+) = g(a)$, then

$$\begin{aligned}A(a) &= \frac{1}{2}(f(a^+) - f(a^-))\mu_g(\{a\}) + \frac{1}{2}(g(a^+) - g(a^-))\mu_f(\{a\}) \\ &= \mu_f(\{a\})\mu_g(\{a\}).\end{aligned}$$

(iii) If f and g are both continuous on the left at a, so that $f(a^-) = f(a)$ and $g(a^-) = g(a)$, then

$$\begin{aligned}A(a) &= \frac{1}{2}(f(a^-) - f(a^+))\mu_g(\{a\}) + \frac{1}{2}(g(a^-) - g(a^+))\mu_f(\{a\}) \\ &= -\mu_f(\{a\})\mu_g(\{a\}).\end{aligned}$$

It is now easy to see why the general theorem for integration by parts takes the form given below. Note that if f and g are functions of bounded variation, then by Theorems 2.7.2 and 2.4.3 the set of points of discontinuity of f and g is either empty or countably infinite.

Theorem 6.2.2 (Integration by Parts)
Let $f, g : I \to \mathbb{R}$ be functions of bounded variation, and let S denote the set of points at which f and g are both discontinuous. Then

$$\int_I f\,dg + \int_I g\,df = \mu_{fg}(I) + \sum_{a \in S} A(a),$$

where

$$\begin{aligned}A(a) = \Big[f(a) &- \frac{1}{2}(f(a^+) + f(a^-))\Big]\mu_g(\{a\}) \\ &+ \Big[g(a) - \frac{1}{2}(g(a^+) + g(a^-))\Big]\mu_f(\{a\}).\end{aligned}$$

In particular,

(i) If S is empty, or if $f(a) = \frac{1}{2}(f(a^+) + f(a^-))$ and $g(a) = \frac{1}{2}(g(a^+) + g(a^-))$ for all $a \in S$, then

$$\int_I f\,dg + \int_I g\,df = \mu_{fg}(I).$$

(ii) If f and g are continuous on the right at all points of S, then

$$\int_I f\,dg + \int_I g\,df = \mu_{fg}(I) + \sum_{a \in S} \mu_f(\{a\})\mu_g(\{a\}).$$

(iii) If f and g are continuous on the left at all points of S, then

$$\int_I f dg + \int_I g df = \mu_{fg}(I) - \sum_{a \in S} \mu_f(\{a\})\mu_g(\{a\}).$$

Example 6-2-1:
Consider the integral $\int_{[0,3)} x^2 \, d\alpha$ that was discussed in Example 6-1-4. Using integration by parts, bearing in mind that x^2 has no points of discontinuity, we have

$$\begin{aligned}\int_{[0,3)} x^2 \, d\alpha + \int_{[0,3)} \alpha \, d(x^2) &= \mu_{x^2\alpha}([0,3)) \\ &= (x^2\alpha)(3^-) - (x^2\alpha)(0^-) \\ &= 9\alpha(3^-) - 0 = 9(5) = 45.\end{aligned}$$

Thus,

$$\begin{aligned}\int_{[0,3)} x^2 \, d\alpha &= 45 - \int_{[0,3)} \alpha \, d(x^2) \\ &= 45 - \int_{(0,3)} \alpha \, d(x^2) \quad \text{by Theorem 6.1.3} \\ &= 45 - \int_{(0,3)} \alpha(2x) \, dx \text{ by Theorem 6.1.7} \\ &= 45 - \int_0^3 2x\alpha \, dx \text{ by Theorem 6.1.3 (a Riemann integral)} \\ &= 45 - \left[\int_1^2 2x(x^2 - 2x + 2) \, dx + \int_2^3 2x(x+2) \, dx\right] \\ &= 45 - \left[\frac{x^4}{2} - \frac{4x^3}{3} + 2x^2\right]_1^2 - \left[\frac{2x^3}{3} + 2x^2\right]_2^3 \\ &= 45 - \left[8 - \frac{32}{3} + 8 - \left(\frac{1}{2} - \frac{4}{3} + 2\right)\right] \\ &\quad - \left[18 + 18 - \left(\frac{16}{3} + 8\right)\right] \\ &= \frac{109}{6},\end{aligned}$$

as before.

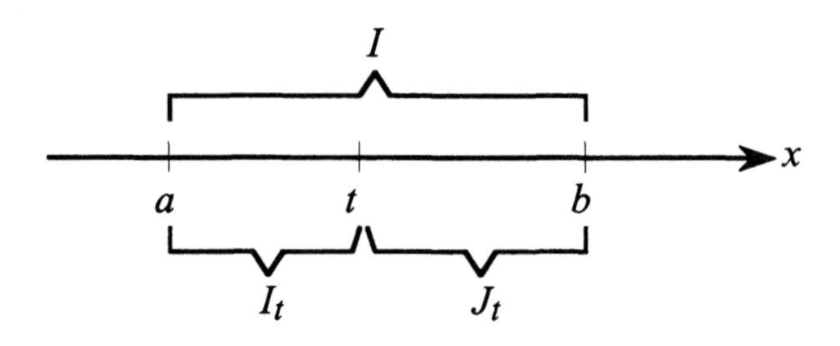

FIGURE 6.8

Exercises 6-2:

1. Use integration by parts to evaluate the integrals in Example 6-1-1 and Exercises 6-1, No. 2(b)(i).
2. (a) Investigate what happens when integration by parts is tried as a method for evaluating the integral in Example 6-1-2. Which hypothesis of Theorem 6.2.2 does not hold in this case?

   (b) Use integration by parts to evaluate $\int_{[0,\infty)} e^{-x}\,d\alpha$, where α is as defined in Example 6-1-2. Check your answer by evaluating the integral directly.

6.3 Integration and Differentiation

We now examine some connections between differentiation and integration. Here we restrict ourselves to the Lebesgue rather than the Lebesgue–Stieltjes integral; that is to say, integration is with respect to x throughout.

For the first theorem, we need some notation. Let I be any interval with endpoints a, b and let t be any real number such that $a < t < b$. We denote by I_t, J_t the intervals defined by $I_t = \{x : x \in I, x \leq t\}$, $J_t = \{x : x \in I, t \leq x\}$ (cf. Figure 6.8).

Theorem 6.3.1 (Fundamental Theorem of Calculus)

(i) If $f : I \to \mathbb{R}$ is integrable over I, then both $F(t) = \int_{I_t} f\,dx$ and $G(t) = \int_{J_t} f\,dx$ are absolutely continuous on I and differentiable a.e. on I, and $F'(t) = f(t)$, $G'(t) = -f(t)$ a.e. on I. If, in addition, f is continuous on I, then "a.e." can be replaced by "everywhere" in the preceding statement.

(ii) *If $F : I \to \mathbb{R}$ is absolutely continuous on I, then it is differentiable a.e. on I, F' is integrable over I, and $F(t) = \int_{I_t} F'\,dx + C$ for all $t \in I$, where C is constant on I.*

Example 6-3-1:

The *error function* erf(t) is defined as an ordinary Riemann integral

$$\operatorname{erf}(t) = \frac{2}{\sqrt{\pi}} \int_0^t e^{-x^2}\,dx$$

for all $t \geq 0$. It is important in statistics (in connection with the normal distribution), and it also arises in the context of certain partial differential equations connected with heat flow. The *complementary error function* erfc(t) is defined by

$$\operatorname{erfc}(t) = \frac{2}{\sqrt{\pi}} \int_t^\infty e^{-x^2}\,dx$$

for all $t \geq 0$. It follows at once from Theorem 6.3.1(a) that for all $t \geq 0$,

$$\frac{d}{dt}[\operatorname{erf}(t)] = \frac{2}{\sqrt{\pi}} e^{-t^2} \quad \text{and} \quad \frac{d}{dt}[\operatorname{erfc}(t)] = -\frac{2}{\sqrt{\pi}} e^{-t^2}.$$

Example 6-3-2:

Let $\alpha : \mathbb{R} \to \mathbb{R}$ be a probability distribution function. If there exists a function $f : \mathbb{R} \to \mathbb{R}$ such that $f \geq 0$ on $\mathbb{R}$ and $\alpha(t) = \int_{(-\infty,t]} f\,dx$ for all $t \in \mathbb{R}$, then f is call a *density* of α. From Theorem 6.3.1, we know that this happens if and only if α is absolutely continuous, and that in this case $\alpha' = f$ a.e. on $\mathbb{R}$. Clearly, discrete distributions, as defined in Section 4-2, do not have densities, since their distribution functions are discontinuous.

If the functions f and g are both densities of α, then $f = g$ a.e., so in this sense we can say that the density of α, if it exists, is unique.

As an example, take the case of the uniform distribution defined in Example 4-2-1:

$$\alpha(x) = \begin{cases} 0, & \text{if } x \leq A, \\ \frac{x-A}{B-A}, & \text{if } A \leq x \leq B, \\ 1, & \text{if } x > B \end{cases}$$

(Figure 6.9). Here we can take the density f of α to be the derivative of α where it exists (which is everwhere except at A and B), and

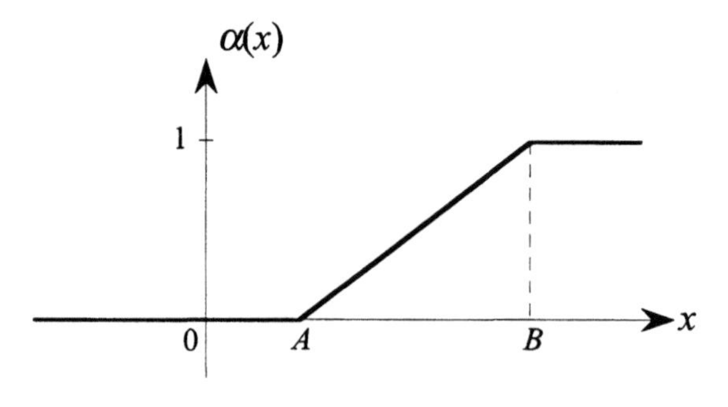

FIGURE 6.9

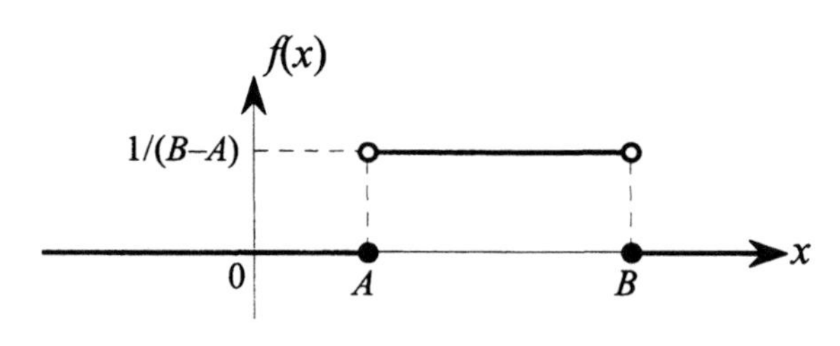

FIGURE 6.10

define f (arbitrarily) to be zero at A and B, giving

$$f(x) = \begin{cases} 0, & \text{if } x \leq A \text{ or } x \geq B, \\ \frac{1}{B-A}, & \text{if } A < x < B \end{cases}$$

(Figure 6.10).

The previous theorem dealt with differentiation of an integral that has a variable interval of integration. It is also important to be able to differentiate functions of the form

$$g(t) = \int_I f(t, x)\, dx,$$

where the variable t appears in the integrand, not in the interval of integration. It is natural to ask whether we can find $g'(t)$ by interchanging the order of the differentiation and integration operations. Thus

$$g'(t) = \frac{d}{dt}\left[\int_I f(t, x)\, dx\right] = \int_I \frac{\partial}{\partial t} f(t, x)\, dx,$$

where in the right-hand integral, x is held constant while differentiation is carried out with respect to t. Simple examples suggest that this is correct. Consider, for example,

$$\begin{aligned} g(t) = \int_0^1 \sin(2t+3x)\,dx &= \left[-\frac{1}{3}\cos(2t+3x)\right]_{x=0}^{x=1} \\ &= -\frac{1}{3}\cos(2t+3) + \frac{1}{3}\cos 2t. \end{aligned}$$

By direct differentiation, $g'(t) = \frac{2}{3}\sin(2t+3) - \frac{2}{3}\sin 2t$. On the other hand,

$$\begin{aligned} \int_0^1 \frac{\partial}{\partial t}\sin(2t+3x)\,dx &= \int_0^1 2\cos(2t+3x)dx \\ &= \left[\frac{2}{3}\sin(2t+3x)\right]_{x=0}^{x=1} \\ &= \frac{2}{3}\sin(2t+3) - \frac{2}{3}\sin 2t, \\ &= g'(t) \end{aligned}$$

as expected. The general theorem, which tells us that this process, called "differentiation under the integral," is legitimate, is as follows:

Theorem 6.3.2 (Differentiation Under the Integral)
Let I and J be any intervals. Let the real-valued function $f(t,x)$ be such that

(i) *$f(t,x)$ is defined for all $t \in J$, $x \in I$;*
(ii) *$f(t,x)$ is integrable with respect to x over I;*
(iii) *For each $t \in J$, $\frac{\partial}{\partial t}f(t,x)$ exists a.e. on I;*
(iv) *For each closed subinterval $J^* \subseteq J$, there exists a function $\lambda : I \to \mathbb{R}$ such that λ is integrable over I, and $|\frac{\partial}{\partial t}f(t,x)| \le \lambda(x)$ for all $x \in I$ and $t \in J^*$.*

Then for each $t \in J$, $\partial/\partial t f(t,x)$ is integrable with respect to x over I, and

$$\frac{d}{dt}\left[\int_I f(t,x)\,dx\right] = \int_I \frac{\partial}{\partial t} f(t,x)\,dx.$$

Example 6-3-3:
Consider the function $g(t) = \int_0^\pi \ln(1+t\cos x)\,dx$, where $-1 < t < 1$. Note that since $|\cos x| \le 1$ for all x, we have that $|t\cos x| \le |t| < 1$

for all $t \in (-1,1)$. Thus $1 + t\cos x > 0$ for all x and all $t \in (-1,1)$, and so both $\ln(1 + t\cos x)$ and

$$\frac{\partial}{\partial t}\ln(1+t\cos x) = \frac{\cos x}{1+t\cos x}$$

are continuous for all x and all $t \in (-1,1)$.

Take any closed subinterval $[a,b] \subseteq (-1,1)$, and let $k = \max\{|a|,|b|\}$. We have that for all x and all $t \in [a,b]$, $|t\cos x| \le k < 1$, and so

$$|1 + t\cos x| \ge 1 - |t\cos x| \ge 1 - k > 0.$$

Thus

$$\left|\frac{\cos x}{1+t\cos x}\right| \le \frac{1}{1-k}$$

for all x and all $t \in [a,b]$. Hence all the conditions of Theorem 6.3.2 are satisfied in this case, where $f(t,x) = \ln(1+t\cos x)$, $I = [0,\pi]$, $J = (-1,1)$, and $\lambda(x) = 1/(1-k)$ for each $J^* = [a,b] \subseteq J$. By Theorem 6.3.2 we have therefore

$$g'(t) = \int_0^\pi \frac{\partial}{\partial t}\ln(1+t\cos x)\,dx = \int_0^\pi \frac{\cos x}{1+t\cos x}\,dx$$

for all $t \in (-1,1)$. Note first that

$$g'(0) = \int_0^\pi \cos x\,dx = \big[\sin x\big]_0^\pi = 0.$$

If $t \in (-1,1)$ is not zero, then

$$\begin{aligned} g'(t) &= \int_0^\pi \frac{1}{t}\left(\frac{1+t\cos x - 1}{1+t\cos x}\right)dx \\ &= \frac{1}{t}\int_0^\pi \left(1 - \frac{1}{1+t\cos x}\right)dx \\ &= \frac{1}{t}\left(\pi - \left[\frac{2}{\sqrt{1-t^2}}\arctan\left(\frac{(1-t)\tan(x/2)}{\sqrt{1-t^2}}\right)\right]_{x=0}^{x=\pi}\right). \end{aligned}$$

(The final equality is not obvious, but it can be verified using any reasonably comprehensive table of integrals.) Now, as $x \to \pi^-$, $\tan(x/2) \to \infty$, and so

$$\frac{(1-t)\tan(x/2)}{\sqrt{1-t^2}} \to \infty$$

as $x \to \pi^-$, and thus

$$\arctan\left(\frac{(1-t)\tan(x/2)}{\sqrt{1-t^2}}\right) \to \frac{\pi}{2}$$

as $x \to \pi^-$. Also, $\arctan(0) = 0$, and so

$$g'(t) = \frac{1}{t}\left(\pi - \frac{\pi}{\sqrt{1-t^2}}\right)$$

for $t \neq 0$, $t \in (-1, 1)$. Thus we have finally

$$g'(x) = \begin{cases} \frac{\pi}{t}\left(1 - \frac{1}{\sqrt{1-t^2}}\right), & \text{if } t \in (-1,1) \text{ and } t \neq 0, \\ 0, & \text{if } t = 0. \end{cases}$$

Since $g(0) = \int_0^\pi (\ln 1)\,dx = \int_0^\pi 0\,dx = 0$, we have that for all $t \in (-1, 1)$

$$g(t) = g(t) - g(0) = \int_0^t g'(x)\,dx$$

by the fundamental theorem of calculus; thus,

$$g(t) = \int_0^t \frac{\pi}{x}\left(1 - \frac{1}{\sqrt{1-x^2}}\right)\,dx.$$

Now (using a table of integrals) we have that

$$\begin{aligned} \int \left(\frac{1}{x} - \frac{1}{x\sqrt{1-x^2}}\right)\,dx &= \ln|x| + \ln\left|\frac{1+\sqrt{1-x^2}}{x}\right| + C \\ &= \ln\left|\frac{x(1+\sqrt{1-x^2})}{x}\right| + C \\ &= \ln\left(1+\sqrt{1-x^2}\right) + C \end{aligned}$$

(note that $1 + \sqrt{1-x^2} > 0$ for $|x| < 1$). Hence,

$$g(t) = \pi\left[\ln(1+\sqrt{1-x^2})\right]_0^t = \pi\left(\ln(1+\sqrt{1-t^2}) - \ln 2\right),$$

and so finally

$$\int_0^\pi \ln(1 + t\cos x)\,dx = \pi \ln\left(\frac{1+\sqrt{1-t^2}}{2}\right)$$

for $-1 < t < 1$.

In the most general case, a function defined by an integral may have the variable appearing both in the limits of integration and in the integrand, for example

$$g(t) = \int_t^{t^2} e^{tx}\,dx.$$

Such a function can be differentiated by using a combination of the fundamental theorem of calculus and differentiation under the integral.

Consider an integral of the form

$$g(t) = \int_{I(t)} f(w(t), x)\,dx,$$

where $I(t)$ is the closed interval $[u(t), v(t)]$. Here we assume that for some interval I_0, $u, v, w : I_0 \to \mathbb{R}$ are differentiable functions of t such that for some interval I_1 we have $w(t) \in I_1$ for all $t \in I_0$. We assume also that for each $t \in I_0$, $f(w, x)$ satisfies the conditions of Theorem 6.3.2 for $w \in I_1$ and $x \in I(t)$. Let $h(u, v, w) = \int_{[u,v]} f(w, x)\,dx$. Then $g(t) = h(u(t), v(t), w(t))$, so by the chain rule

$$\begin{aligned} g'(t) &= \frac{\partial h}{\partial u}\frac{du}{dt} + \frac{\partial h}{\partial v}\frac{dv}{dt} + \frac{\partial h}{\partial w}\frac{dw}{dt} \\ &= -f(w, u)\frac{du}{dt} + f(w, v)\frac{dv}{dt} + \left(\int_{[u,v]} \frac{\partial}{\partial w} f(w, x)\,dx\right)\frac{dw}{dt}, \end{aligned}$$

where the fundamental theorem of calculus has been used to find $\partial h/\partial u$ and $\partial h/\partial v$, and differentiation under the integral to find $\partial h/\partial w$. In particular, if we take $w(t) = t$ for $t \in I_0$, then $dw/dt = 1$, and we obtain the following result, often referred to as **Leibniz's rule**:

$$\begin{aligned} \frac{d}{dt}\left[\int_{[u(t),v(t)]} f(t, x)\,dx\right] &= -f(t, u(t))\frac{du}{dt} + f(t, v(t))\frac{dv}{dt} \\ &\quad + \int_{[u(t),v(t)]} \frac{\partial}{\partial t} f(t, x)\,dx. \end{aligned} \tag{6.1}$$

If f happens to be independent of t, then we obtain the important special case

$$\frac{d}{dt}\left[\int_{[u(t),v(t)]} f(x)\,dx\right] = -f(u(t))\frac{du}{dt} + f(v(t))\frac{dv}{dt}. \tag{6.2}$$

Consider the function $g(t) = \int_t^{t^2} e^{tx}\,dx$ for $t \geq 1$. Here $I_0 = [1, \infty)$, and since $w(t) = t$, we have also $I_1 = [1, \infty)$. Take any $t \geq 1$, and let $[a, b]$ be any subinterval of $[1, \infty)$. Then

$$\left|\frac{\partial}{\partial w} e^{wx}\right| = xe^{wx} \leq xe^{bx}$$

for all $w \in [a, b]$ and $x \in [t, t^2]$, so we can apply Leibniz's rule to obtain

$$\begin{aligned}
g'(t) &= -e^{t(t)}\frac{d}{dt}(t) + e^{t(t^2)}\frac{d}{dt}(t^2) + \int_t^{t^2} xe^{tx}\,dx \\
&= -e^{t^2} + 2te^{t^3} + \left\{\left[\frac{xe^{tx}}{t}\right]_t^{t^2} - \int_t^{t^2} \frac{e^{tx}}{t}\,dx\right\} \quad \text{integrating by parts} \\
&= -e^{t^2} + 2te^{t^3} + \left\{\left[te^{t^3} - e^{t^2}\right] - \left[\frac{e^{tx}}{t^2}\right]_t^{t^2}\right\} \\
&= -e^{t^2} + 2te^{t^3} + \left\{te^{t^3} - e^{t^2} - \left[\frac{e^{t^3}}{t^2} - \frac{e^{t^2}}{t^2}\right]\right\} \\
&= 3te^{t^3} - 2e^{t^2} - \frac{1}{t^2}\left(e^{t^3} - e^{t^2}\right).
\end{aligned}$$

In this particular case we can check by evaluating $g(t)$ directly:

$$g(t) = \left[\frac{e^{tx}}{t}\right]_t^{t^2} = \frac{1}{t}\left(e^{t^3} - e^{t^2}\right);$$

thus,

$$\begin{aligned}
g'(t) &= \frac{1}{t}\left(3t^2e^{t^3} - 2te^{t^2}\right) - \frac{1}{t^2}\left(e^{t^3} - e^{t^2}\right) \\
&= 3te^{t^3} - 2e^{t^2} - \frac{1}{t^2}\left(e^{t^3} - e^{t^2}\right),
\end{aligned}$$

as before.

Example 6-3-4:
Consider the differential equation

$$\frac{d^2y}{dt^2} = g(t),$$

with initial conditions $y(0) = c_1$, $y'(0) = c_2$. Assume that g is continuous on the interval $[0, \infty)$. For all $t \geq 0$, define

$$y(t) = c_1 + c_2 t + \int_0^t (t - x) g(x)\, dx.$$

Clearly, $y(0) = c_1$. The function $f(w, x) = (w - x)g(x)$ certainly satisfies conditions (i) and (ii) of Theorem 6.3.2, for $w \in [0, \infty)$ and $x \in [0, t]$ $(t \geq 0)$. Further, for each $t \in [0, \infty)$ we have that

$$\left| \frac{\partial}{\partial w} f(w, x) \right| = |g(x)|$$

(independent of w) is continuous and therefore integrable over $[0, t]$. It follows by Leibniz's rule that

$$\begin{aligned} y'(t) &= c_2 + (t - t)g(t) + \int_0^t g(x)\, dx \\ &= c_2 + \int_0^t g(x)\, dx. \end{aligned}$$

Thus $y'(0) = c_2$; also, we have by the fundamental theorem of calculus that

$$y''(t) = g(t),$$

and so the function $y(t)$ defined above is the solution of the given initial value problem.

As a final comment on the relationship between the integral and the derivative, we point out a serious gap in the Lebesgue theory. Recall that an antiderivative (sometimes called an indefinite integral) of a function f is a function F such that $F' = f$. It turns out that there exist functions that are "integrable" in the sense of having an antiderivative at all points of a certain interval but are not Lebesgue integrable on that interval. An example of such a function is given in Section 10.1, where we revisit this matter.

Exercises 6-3:

1. Use L'Hospital's rule to find

$$\text{(a)} \lim_{t \to 0^+} \frac{\operatorname{erf}(t)}{t} \quad \text{and} \quad \text{(b)} \lim_{t \to \infty} t \operatorname{erfc}(t).$$

2. By writing erf(x) as $1 \times \text{erf}(x)$ and integrating by parts, show that
$$\int_0^t \text{erf}(x)\,dx = t\,\text{erf}(t) - \frac{1}{\sqrt{\pi}}(1 - e^{-t^2}).$$
3. Use a table of integrals to show that
$$\int_0^\pi \frac{1}{t - \cos x}\,dx = \frac{\pi}{\sqrt{t^2 - 1}}$$
for all $t > 1$. By differentiating both sides of this equation with respect to t, evaluate
$$\int_0^\pi \frac{1}{(t - \cos x)^2}\,dx.$$
4. Given that $g(t) = \int_0^t \sin(x-t)\,dx$, find $g'(t)$ by using Leibniz's rule. Check by evaluating $g(t)$ directly and then differentiating.
5. Find $g'(t)$ if
$$g(t) = \int_t^{t^2} \frac{1}{x}\sin(tx)\,dx,$$
where $t \geq 1$.
6. Assuming that g is continuous on the interval $[0, \infty)$, show that the function
$$y(t) = \frac{1}{k}c_2 \sin(kt) + c_1 \cos(kt) + \frac{1}{k}\int_0^t g(x)\sin\left\{k(t-x)\right\}dx,$$
for $t \geq 0$, satisfies the differential equation
$$\frac{d^2y}{dt^2} + k^2 y = g(t),$$
where $k > 0$, together with the initial conditions $y(0) = c_1$, $y'(0) = c_2$.
7. Assuming that g is continuous on the interval $[0, \infty)$, show that the function
$$y(t) = \frac{1}{n!}\int_0^t (t-x)^n g(x)\,dx$$
satisfies the differential equation
$$\frac{d^{n+1}y}{dt^{n+1}} = g(t)$$
together with the initial conditions
$$y(0) = 0,\ \ y^{(1)}(0) = 0, \ldots, y^{(n)}(0) = 0.$$

Double and Repeated Integrals

Lebesgue–Stieltjes integrals of functions of more than one variable can be defined using the same approach as was used in Section 4.5 for functions of one variable. For the sake of simplicity we will discuss only functions of two variables. The process for functions of more than two variables is completely analogous.

7.1 Measure of a Rectangle

We define a **rectangle** to be a set of the form $I_1 \times I_2 \subseteq \mathbb{R}^2$, where I_1 and I_2 are intervals. For monotone increasing functions $\alpha_1, \alpha_2 : \mathbb{R} \to \mathbb{R}$ we define the $\alpha_1 \times \alpha_2$**-measure** of $I_1 \times I_2$, denoted by $\mu_{\alpha_1 \times \alpha_2}(I_1 \times I_2)$, by

$$\mu_{\alpha_1 \times \alpha_2}(I_1 \times I_2) = \mu_{\alpha_1}(I_1) \times \mu_{\alpha_2}(I_2).$$

For example, if α_1 and α_2 are the functions defined in Exercises 4-1, problems 1 and 2, respectively, then

$$\begin{aligned} &\mu_{\alpha_1}((0,1)) = 1 - e^{-1}, \quad &&\mu_{\alpha_1}([0,1)) = 3 - e^{-1}, \\ &\mu_{\alpha_2}((0,1)) = 0, &&\mu_{\alpha_2}([0,1)) = 1, \end{aligned}$$

and therefore

$$\mu_{\alpha_1\times\alpha_2}((0,1)\times(0,1)) = 0, \quad \mu_{\alpha_1\times\alpha_2}((0,1)\times[0,1)) = 1 - e^{-1},$$
$$\mu_{\alpha_1\times\alpha_2}([0,1)\times[0,1)) = 3 - e^{-1}.$$

7.2 Simple Sets and Simple Functions in Two Dimensions

A **simple set** in $\mathbb{R}^2$ is a subset of $\mathbb{R}^2$ that can be expressed as the union of a finite collection of disjoint rectangles. Just as for simple sets in $\mathbb{R}$, we can define the measure of a simple set in $\mathbb{R}^2$. If $\alpha_1, \alpha_2 : \mathbb{R} \to \mathbb{R}$ are monotone increasing functions and S is a simple set of the form

$$S = \bigcup_{j=1}^{m} (I_{1,j} \times I_{2,j}),$$

where $I_{1,1} \times I_{2,1}, I_{1,2} \times I_{2,2}, \ldots, I_{1,m} \times I_{2,m}$ are pairwise disjoint rectangles, then the $\alpha_1 \times \alpha_2$-measure of S is defined by

$$\mu_{\alpha_1\times\alpha_2}(S) = \sum_{j=1}^{m} \mu_{\alpha_1\times\alpha_2}(I_{1,j} \times I_{2,j}).$$

The properties of simple sets in $\mathbb{R}^2$ and their measures are the same as those described in Section 5-3 for simple sets in $\mathbb{R}$.

We can now define simple functions of two variables by analogy with step functions of one variable (see Sections 2-5 and 4-4). We could continue to use the term "step functions," but customary usage restricts this term to functions of one variable.

A function $\theta : \mathbb{R}^2 \to \mathbb{R}$ is a **simple function** if there is a simple set

$$S = \bigcup_{j=1}^{n} (I_{1,j} \times I_{2,j})$$

and a list $(c_1, c_2, \ldots, c_n)$ of finite, nonzero real numbers such that

$$\theta(x,y) = \begin{cases} c_j, & \text{if } (x,y) \in I_{1,j} \times I_{2,j}, \quad (j = 1, 2, \ldots, n), \\ 0, & \text{if } (x,y) \in \mathbb{R}^2 - S. \end{cases}$$

The set S is called the **support** of θ. The properties of step functions given in Section 2-5 carry over without difficulty to simple functions. If $\alpha_1, \alpha_2 : \mathbb{R} \to \mathbb{R}$ are monotone increasing functions, we define the generalized "volume" $A_{\alpha_1 \times \alpha_2}(\theta)$ in a way exactly analogous to the definition of $A_\alpha(\theta)$ in Section 4-4.

7.3 The Lebesgue–Stieltjes Double Integral

Let S be a subset of $\mathbb{R}^2$ and let $f : S \to \mathbb{R}$ be a function. We extend the definition of f to $\mathbb{R}^2$ by defining $f(x, y)$ to be zero if $(x, y) \in \mathbb{R}^2 - S$. Let $\alpha_1, \alpha_2 : \mathbb{R} \to \mathbb{R}$ be monotone increasing functions. The Lebesgue–Stieltjes **double integral** of f, denoted by

$$\int_{\mathbb{R}^2}\!\!\int f \, d(\alpha_1 \times \alpha_2),$$

is defined by a process that is almost word-for-word the same as that used for the single-variable integral in Section 4-5. The only change is that α-summable step functions on I are replaced by $\alpha_1 \times \alpha_2$-summable simple functions on $\mathbb{R}^2$. All the elementary properties analogous to those proved in Sections 5-1 and 5-2 carry over and are proved in the same way, and the same goes for the convergence theorems of Section 5-3 and the definitions of measurable functions and measurable sets given in Section 5-4.

In practice, the evaluation of double integrals is invariably done, as in elementary calculus, by converting them to repeated integrals.

7.4 Repeated Integrals and Fubini's Theorem

Let $f : \mathbb{R}^2 \to \mathbb{R}$ be a function. For any $y \in \mathbb{R}$ we define the single-variable function $f(\cdot, y) : x \to f(x, y)$, and for any $x \in \mathbb{R}$ we likewise define the function $f(x, \cdot) : y \to f(x, y)$. Let α_1, α_2 be monotone increasing functions. If for each $y \in \mathbb{R}$, $\int_{\mathbb{R}} f(\cdot, y)\, d\alpha_1$ exists, then this defines a function $f_2 : y \to \int_{\mathbb{R}} f(\cdot, y)\, d\alpha_1$. If $\int_{\mathbb{R}} f_2\, d\alpha_2$ exists, we call this a **repeated integral** of f and write it as $\int_{\mathbb{R}} \int_{\mathbb{R}} f\, d\alpha_1\, d\alpha_2$. If for each $x \in \mathbb{R}$, $\int_{\mathbb{R}} f(x, \cdot)\, d\alpha_2$ exists, we define $f_1 : x \to \int_{\mathbb{R}} f(x, \cdot)\, d\alpha_2$, and if $\int_{\mathbb{R}} f_1\, d\alpha_1$ exists, it gives us the repeated integral of f with the opposite order of integration, written $\int_{\mathbb{R}} \int_{\mathbb{R}} f\, d\alpha_2\, d\alpha_1$. In most cases calculation shows that $\int_{\mathbb{R}} \int_{\mathbb{R}} f\, d\alpha_1\, d\alpha_2$ and $\int_{\mathbb{R}} \int_{\mathbb{R}} f\, d\alpha_2\, d\alpha_1$ have the same value, but this is not always the case. Consider, for example, the improper Riemann repeated integrals

$$\int_0^1 \int_0^1 \frac{x-y}{(x+y)^3}\, dx\, dy \quad \text{and} \quad \int_0^1 \int_0^1 \frac{x-y}{(x+y)^3}\, dy\, dx.$$

We have

$$\begin{aligned}
\int_0^1 \int_0^1 \frac{x-y}{(x+y)^3}\, dx\, dy &= \int_0^1 \int_0^1 \left(\frac{1}{(x+y)^2} - \frac{2y}{(x+y)^3} \right) dx\, dy \\
&= \int_0^1 \left[-\frac{1}{(x+y)} + \frac{y}{(x+y)^2} \right]_{x=0}^{x=1} dy \\
&= \int_0^1 \left(-\frac{1}{1+y} + \frac{y}{(1+y)^2} + \frac{1}{y} - \frac{1}{y} \right) dy \\
&= \int_0^1 \frac{-1}{(1+y)^2}\, dy \\
&= \left[\frac{1}{1+y} \right]_0^1 = -\frac{1}{2}.
\end{aligned}$$

However,

$$\begin{aligned}
\int_0^1 \int_0^1 \frac{x-y}{(x+y)^3}\, dy\, dx &= \int_0^1 \int_0^1 \frac{y-x}{(x+y)^3}\, dx\, dy \text{ (interchanging } x \text{ and } y\text{)} \\
&= -\int_0^1 \int_0^1 \frac{x-y}{(x+y)^3}\, dx\, dy \\
&= \frac{1}{2},
\end{aligned}$$

and so the two repeated integrals have different values.

It turns out, though we shall not prove it, that this cannot happen if either repeated integral is absolutely convergent. We can easily verify that this condition does not hold in our example. We

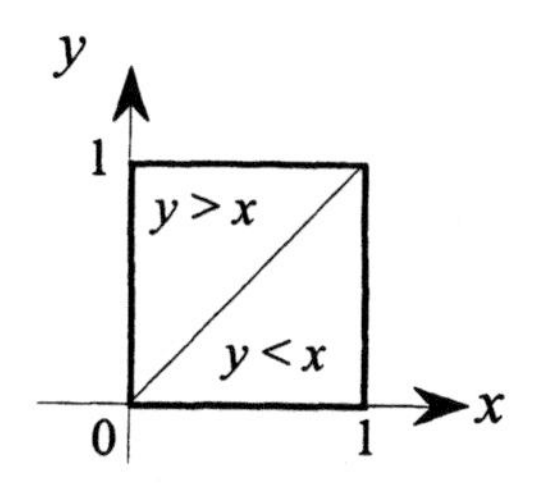

FIGURE 7.1

investigate

$$\int_0^1 \int_0^1 \frac{|x-y|}{(x+y)^3}\, dx\, dy$$

by splitting the region of integration into two parts, one where $y < x$ and one where $y > x$ (cf. Figure 7.1). Then,

$$\begin{aligned}
\int_0^1 \int_0^1 \frac{|x-y|}{(x+y)^3}\, dx\, dy &= \int_0^1 \int_y^1 \frac{x-y}{(x+y)^3}\, dx\, dy + \int_0^1 \int_0^y \frac{y-x}{(x+y)^3}\, dx\, dy \\
&= \int_0^1 \left\{ \int_y^1 \left(\frac{1}{(x+y)^2} - \frac{2y}{(x+y)^3} \right) dx \right. \\
&\qquad \left. - \int_0^y \left(\frac{1}{(x+y)^2} - \frac{2y}{(x+y)^3} \right) dx \right\} dy \\
&= \int_0^1 \left\{ \left[-\frac{1}{x+y} + \frac{y}{(x+y)^2} \right]_{x=y}^{x=1} \right. \\
&\qquad \left. - \left[-\frac{1}{x+y} + \frac{y}{(x+y)^2} \right]_{x=0}^{x=y} \right\} dy \\
&= \int_0^1 \left\{ \frac{-1}{1+y} + \frac{y}{(1+y)^2} + \frac{1}{2y} \right. \\
&\qquad \left. - \frac{1}{4y} - \left(-\frac{1}{2y} + \frac{1}{4y} + \frac{1}{y} - \frac{1}{y} \right) \right\} dy \\
&= \int_0^1 \left(-\frac{1}{(1+y)^2} + \frac{1}{2y} \right) dy \\
&= -\frac{1}{2} + \frac{1}{2} \int_0^1 \frac{1}{y}\, dy,
\end{aligned}$$

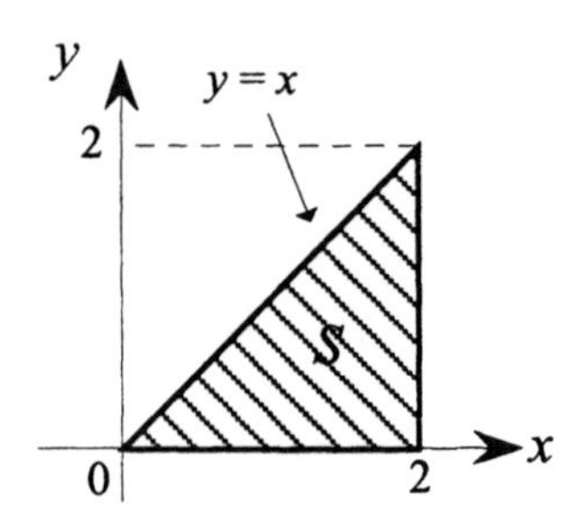

FIGURE 7.2

and so this integral (and likewise $\int_0^1 \int_0^1 \frac{|x-y|}{(x+y)^3}\, dy\, dx$) does not converge.

The fundamental theorem that relates double and repeated integrals using the absolute convergence condition is called Fubini's theorem:

Theorem 7.4.1 (Fubini's Theorem)
If $\alpha_1, \alpha_2 : \mathbb{R} \to \mathbb{R}$ are monotone increasing functions and $f : \mathbb{R}^2 \to \mathbb{R}$ is $\alpha_1 \times \alpha_2$-measurable, then the existence of any one of the integrals

$$\int_{\mathbb{R}^2}\int |f|\, d(\alpha_1 \times \alpha_2), \quad \int_{\mathbb{R}}\int_{\mathbb{R}} |f|\, d\alpha_1\, d\alpha_2, \quad \int_{\mathbb{R}}\int_{\mathbb{R}} |f|\, d\alpha_2\, d\alpha_1,$$

implies the existence and equality of the integrals

$$\int_{\mathbb{R}^2}\int f\, d(\alpha_1 \times \alpha_2), \quad \int_{\mathbb{R}}\int_{\mathbb{R}} f\, d\alpha_1\, d\alpha_2, \quad \int_{\mathbb{R}}\int_{\mathbb{R}} f\, d\alpha_2\, d\alpha_1.$$

In practice, the functions that arise are almost always measurable, so Fubini's theorem justifies the use of repeated integrals to evaluate double integrals, provided that one of the repeated integrals is absolutely convergent. The details can be very messy, so we will confine ourselves to one example.

Example 7-4-1:
Let $S = \{(x, y) : 0 \leq x \leq 2,\ 0 \leq y \leq x\}$ (see Figure 7.2) and let $f : \mathbb{R}^2 \to \mathbb{R}$ be defined by

$$f(x, y) = \begin{cases} 1 + xy, & \text{if } (x, y) \in S, \\ 0, & \text{if } (x, y) \in \mathbb{R}^2 - S. \end{cases}$$

Let α_1 and α_2 be the functions defined in Exercises 4-1, problems 1 and 2, respectively. We will evaluate $\int_{\mathbb{R}^2}\int f\,d(\alpha_1 \times \alpha_2)$ by evaluating the repeated integral $\int_{\mathbb{R}}\int_{\mathbb{R}} f\,d\alpha_1\,d\alpha_2$, and check by evaluating $\int_{\mathbb{R}}\int_{\mathbb{R}} f\,d\alpha_2\,d\alpha_1$. Since the integrand is nonnegative, our procedure will be justified (by Fubini's theorem) if one of the repeated integrals exists.

Evaluation of $\int_{\mathbb{R}}\int_{\mathbb{R}} f\,d\alpha_1\,d\alpha_2$:

We have that

$$f(\cdot, y)(x) = \begin{cases} 1 + xy, & \text{if } 0 \le y \le x \le 2, \\ 0, & \text{otherwise;} \end{cases}$$

hence,

$$\begin{aligned} f_2(y) &= \int_{\mathbb{R}} f(\cdot, y)\,d\alpha_1 \\ &= \begin{cases} \int_{[y,2]}(1 + xy)\,d\alpha_1, & \text{if } 0 \le y \le 2, \\ 0, & \text{otherwise} \end{cases} \\ &= \begin{cases} 1(\alpha_1(0^+) - \alpha_1(0^-)) + \int_0^2 1 \frac{d}{dx}(3 - e^{-x})\,dx, & \text{if } y = 0, \\ \int_y^2 (1 + xy)\,\frac{d}{dx}(3 - e^{-x})\,dx, & \text{if } 0 < y \le 2, \\ 0, & \text{otherwise} \end{cases} \\ &= \begin{cases} 2 + \int_0^2 e^{-x}\,dx, & \text{if } y = 0, \\ \int_y^2 (1 + xy)e^{-x}\,dx, & \text{if } 0 < y \le 2, \\ 0, & \text{otherwise} \end{cases} \\ &= \begin{cases} 3 - e^{-2}, & \text{if } y = 0, \\ -e^{-2} - 3ye^{-2} + e^{-y} + y^2e^{-y} + ye^{-y}, & \text{if } 0 < y \le 2, \\ 0, & \text{otherwise.} \end{cases} \end{aligned}$$

Therefore,

$$
\begin{aligned}
\int_{\mathbb{R}}\int_{\mathbb{R}} f\, d\alpha_1\, d\alpha_2 &= \int_{\mathbb{R}} f_2\, d\alpha_2 \\
&= (3-e^{-2})(\alpha_2(0^+)-\alpha_2(0^-)) \\
&\quad + \int_0^1 (-e^{-2}-3ye^{-2}+e^{-y}+y^2e^{-y}+ye^{-y})\frac{d}{dy}(1)dy \\
&\quad +(-4e^{-2}+3e^{-1})(\alpha_2(1^+)-\alpha_2(1^-)) \\
&\quad + \int_1^2 (-e^{-2}-3ye^{-2}+e^{-y}+y^2e^{-y}+ye^{-y})\frac{d}{dy}(4)dy \\
&\quad + 0(\alpha_2(2^+)-\alpha_2(2^-)) \\
&= (3-e^{-2})(1)+(-4e^{-2}+3e^{-1})(3) \\
&= 3+9e^{-1}-13e^{-2}.
\end{aligned}
$$

Evaluation of $\int_{\mathbb{R}}\int_{\mathbb{R}} f\, d\alpha_1\, d\alpha_2$:

We have that

$$
f(x,\cdot)(y) = \begin{cases} 1+xy, & \text{if } 0 \le y \le x \le 2, \\ 0, & \text{otherwise;} \end{cases}
$$

hence,

$$
\begin{aligned}
f_1(x) &= \int_{\mathbb{R}} f(x,\cdot)\, d\alpha_2 \\
&= \begin{cases} \int_{[0,x]}(1+xy)\, d\alpha_2, & \text{if } 0 \le x \le 2\,, \\ 0, & \text{otherwise} \end{cases}
\end{aligned}
$$

$$
= \begin{cases}
1(\alpha_2(0^+) - \alpha_2(0^-)) + \int_0^x (1+xy)\frac{d}{dy}(1)\,dy, & \text{if } 0 \le x < 1, \\
1(\alpha_2(0^+) - \alpha_2(0^-)) + \int_0^1 (1+xy)\frac{d}{dy}(1)\,dy & \\
\quad + (1+x)(\alpha_2(1^+) - \alpha_2(1^-)) + \int_1^x (1+xy)\frac{d}{dy}(4)\,dy, & \\
 & \text{if } 1 \le x < 2, \\
1(\alpha_2(0^+) - \alpha_2(0^-)) + \int_0^1 (1+xy)\frac{d}{dy}(1)\,dy & \\
\quad + (1+x)(\alpha_2(1^+) - \alpha_2(1^-)) + \int_1^2 (1+xy)\frac{d}{dy}(4)\,dy & \\
\quad + (1+2x)(\alpha_2(2^+) - \alpha_2(2^-)), & \text{if } x = 2, \\
0, & \text{otherwise}
\end{cases}
$$

$$
= \begin{cases}
1, & \text{if } 0 \le x < 1, \\
4+3x, & \text{if } 1 \le x < 2, \\
6+7x, & \text{if } x = 2, \\
0, & \text{otherwise.}
\end{cases}
$$

Therefore,

$$
\begin{aligned}
\int_{\mathbb{R}}\int_{\mathbb{R}} f\,d\alpha_2\,d\alpha_1 &= \int_{\mathbb{R}} f_1\,d\alpha_1 \\
&= 1(\alpha_1(0^+) - \alpha_1(0^-)) + \int_0^1 1\frac{d}{dx}(3-e^{-x})\,dx \\
&\quad + \int_1^2 (4+3x)\frac{d}{dx}(3-e^{-x})\,dx \\
&= 2 + \int_0^1 e^{-x}\,dx + \int_1^2 (4+3x)e^{-x}\,dx \\
&= 2 + \left[-e^{-x}\right]_0^1 + \left[-4e^{-x} - 3xe^{-x} - 3e^{-x}\right]_1^2 \\
&= 3 + 9e^{-1} - 13e^{-2} = \int_{\mathbb{R}}\int_{\mathbb{R}} f\,d\alpha_1\,d\alpha_2.
\end{aligned}
$$

Exercise 7-4:
Let $S = \{(x, y) : 0 \leq x \leq 2,\ \ 0 \leq y \leq 2 - x\}$ and let $f : \mathbb{R}^2 \to \mathbb{R}$ be defined by

$$f(x, y) = \begin{cases} e^x \sin y, & \text{if } (x, y) \in S, \\ 0, & \text{if } (x, y) \in \mathbb{R}^2 - S. \end{cases}$$

Let α_1 and α_2 be the functions defined in Exercises 4-1, problems 1 and 2, respectively. Verify that

$$\int_{\mathbb{R}} \int_{\mathbb{R}} f \, d\alpha_1 \, d\alpha_2 = \int_{\mathbb{R}} \int_{\mathbb{R}} f \, d\alpha_2 \, d\alpha_1 .$$

8 CHAPTER The Lebesgue Spaces L^p

There are many mathematical problems for which the solution is a function of some kind, and it is often both possible and convenient to specify in advance the set of functions within which the solution is to be sought. For example, the solution to a first-order differential equation might be specified as being differentiable on the whole real line. The set of functions differentiable on the whole real line has the useful property that sums and constant multiples of functions in the set are also in the set. In fact, this set of functions has the structure of a vector space, where the "vectors" are functions.

Beyond the algebraic properties associated with vector spaces, many problems are solved by use of series or sequences of functions, and it is desirable that any "limits" also be in the set. Some of the most useful sets of functions have this property. We have seen in Section 3-3 that the limit of a sequence of Riemann integrable functions does not necessarily yield a Riemann integrable function, and this signals that the sets defined using the Riemann integral may not be suitable for many applications. In contrast, sets defined using the Lebesgue integral have the desirable "limit properties."

There are a number of sets of functions that are vector spaces that are of importance in subjects such as differential and integral equations, real and complex function theory, and probability theory, along with the fields of applied mathematics where these subjects play a significant role. Some of the most important of these function sets make use of the Lebesgue integral in their definitions, and so it is appropriate to discuss them here.

In this chapter we aim to give the reader an overview of some of these function sets. We neither go into all the technical details nor attempt a comprehensive survey. References are given where more detail can be found if desired.

8.1 Normed Spaces

The reader has probably encountered the concept of a finite-dimensional vector space. These spaces are modeled after the set of vectors in $\mathbb{R}^n$. Vector spaces, however, can be defined more generally and need not be finite-dimensional. Indeed, most the vector spaces of interest in analysis are not finite-dimensional. Recall that a **vector space** is a nonempty set X equipped with the operations of addition '+' and scalar multiplication. For any elements f, g, h in X and any scalars α, β these operations have the following properties:

(i) $f + g \in X$;

(ii) $f + g = g + f$;

(iii) $f + (g + h) = (f + g) + h$;

(iv) there is a unique element $\mathbf{0}$ (called zero) in X such that $f + \mathbf{0} = f$ for all $f \in X$;

(v) for each element $f \in X$ there is a unique element $(-f) \in X$ such that $f + (-f) = \mathbf{0}$;

(vi) $\alpha f \in X$;

(vii) $\alpha(f + g) = \alpha f + \alpha g$;

(viii) $(\alpha + \beta)f = \alpha f + \beta f$;

(ix) $(\alpha\beta)f = \alpha(\beta f)$;

(x) $1 \cdot f = f$.

For our purposes, the scalars will be either real or complex numbers. We shall use the term *complex vector space* if the scalars are complex numbers when there is some danger of confusion.

Example 8-1-1:
The set of vectors $\{(x_1, x_2, \ldots, x_n) : x_k \in \mathbb{R}, k = 1, 2, \ldots, n\}$ is denoted by $\mathbb{R}^n$. Let $\mathbf{x} = (x_1, x_2, \ldots, x_n)$ and $\mathbf{y} = (y_1, y_2, \ldots, y_n)$ be vectors in $\mathbb{R}^n$. If addition is defined by

$$\mathbf{x} + \mathbf{y} = (x_1 + y_1, x_2 + y_2, \ldots, x_n + y_n)$$

and scalar multiplication by

$$\alpha\mathbf{x} = (\alpha x_1, \alpha x_2, \ldots, \alpha x_n)$$

for any $\alpha \in \mathbb{R}$, then $\mathbb{R}^n$ is a vector space.

Similarly, the set $\mathbb{C}^n = \{(z_1, z_2, \ldots, z_n) : z_k \in \mathbb{C}, k = 1, 2, \ldots, n\}$ of complex vectors is a complex vector space when addition is defined by

$$\mathbf{z} + \mathbf{w} = (z_1 + w_1, z_2 + w_2, \ldots, z_n + w_n)$$

for any vectors $\mathbf{z} = (z_1, z_2, \ldots, z_n)$, $\mathbf{w} = (w_1, w_2, \ldots, w_n)$, and scalar multiplication by

$$\alpha\mathbf{x} = (\alpha x_1, \alpha x_2, \ldots, \alpha x_n),$$

where $\alpha \in \mathbb{C}$. The vector spaces $\mathbb{R}^n$ and $\mathbb{C}^n$ are essentially the prototypes for more abstract vector spaces.

Example 8-1-2:
Let $C[a, b]$ denote the set of all functions $f : [a, b] \to \mathbb{R}$ that are continuous on the interval $[a, b]$. If for any $f, g \in C[a, b]$, addition is defined by

$$(f + g)(x) = f(x) + g(x),$$

and scalar multiplication by

$$(\alpha f)(x) = \alpha f(x)$$

for $\alpha \in \mathbb{R}$, then it is not difficult to see that $C[a, b]$ is a vector space.

Example 8-1-3:
Let ℓ^1 denote the set of sequences $\{a_n\}$ in $\mathbb{R}$ such that the series $\sum_{n=1}^{\infty} |a_n|$ is convergent, and define addition so that for any two elements $A = \{a_n\}$, $B = \{b_n\}$,

$$A + B = \{a_n + b_n\},$$

and scalar multiplication so that

$$\alpha A = \{\alpha a_n\}.$$

Then ℓ^1 is also a vector space.

The above examples show that the elements in different vector spaces can be very different in nature. More importantly, however, there is a significant difference between a vector space such as $\mathbb{R}^n$ and one such as $C[a, b]$ having to do with "dimension." The space $\mathbb{R}^n$ has a basis: Any set of n linearly independent vectors in $\mathbb{R}^n$ such as $\mathbf{e}_1 = (1, 0, \ldots, 0), \mathbf{e}_2 = (0, 1, \ldots, 0) \ldots, \mathbf{e}_n = (0, 0, \ldots, 1)$ forms a basis. The concept of dimension is tied to the number of elements in a basis for spaces such as $\mathbb{R}^n$, but what would be a basis for a space like $C[a, b]$? In order to make some progress on this question we need first to define what is meant by a linearly independent set when the set itself might contain an infinite number of elements. We say that a set is **linearly independent** if every *finite* subset is linearly independent; otherwise, it is called **linearly dependent**. If there exists a positive integer n such that a vector space X has n linearly independent vectors but any set of $n + 1$ vectors is linearly dependent, then X is called **finite-dimensional**. If no such integer exists, then X is called **infinite-dimensional**. We will return to the question of bases for certain infinite-dimensional vector spaces in Chapter 9.

A **subspace** of a vector space X is a subset of X that is itself a vector space under the same operations of addition and scalar multiplication. For example, the set of functions $f : [a, b] \to \mathbb{R}$ such that f is differentiable on $[a, b]$ is a subspace of $C[a, b]$. Given any vectors $x_1, x_2, \ldots, x_n$ in a vector space X, a subspace can always be formed by generating all the linear combinations involving the x_k, i.e., all the vectors of the form $\alpha_1 x_1 + \alpha_2 x_2 + \cdots + \alpha_n x_n$, where the α_k's are scalars. Given any finite set $S \subset X$ the subspace of X formed in this manner is called the **span** of S and denoted by $[S]$. If $S \subset X$ has an infinite number of elements, then the span of S is defined to be the set of all *finite* linear combinations of elements of S.

Vector spaces of functions such as $C[a, b]$ are often referred to simply as *function spaces*. After the next section we shall be concerned almost exclusively with function spaces, and to avoid repetition we shall agree here that for any function space the operations of addition and scalar multiplication will be defined pointwise as was done for the space $C[a, b]$ in Example 8-1-2.

Vector spaces are purely algebraic objects, and in order to do any analysis we need to further specialize. In particular, basic concepts such as convergence require some means of measuring the "distance" between objects in the vector space. This leads us to the concept of a norm. A **norm** on a vector space X is a real-valued function on X whose value at $f \in X$ is denoted by $\|f\|$ and that has the following properties:

(i) $\|f\| \geq 0$;

(ii) $\|f\| = 0$ if and only if $f = \mathbf{0}$;

(iii) $\|\alpha f\| = |\alpha| \|f\|$;

(iv) $\|f + g\| \leq \|f\| + \|g\|$ (the triangle inequality).

Here, f and g are arbitrary elements in X, and α is any scalar. A vector space X equipped with a norm $\|\cdot\|$ is called a **normed vector space**.

Example 8-1-4:
For any $\mathbf{x} \in \mathbb{R}^n$ let $\|\cdot\|_e$ be defined by

$$\|\mathbf{x}\|_e = \{(x_1^2 + (x_2)^2 + \cdots + (x_n)^2\}^{1/2}.$$

Then $\|\cdot\|_e$ is a norm on $\mathbb{R}^n$. This function is called the *Euclidean* norm on $\mathbb{R}^n$. Another norm on $\mathbb{R}^n$ is given by

$$\|\mathbf{x}\|_T = |x_1| + |x_2| + \cdots + |x_n|.$$

Example 8-1-5:
The function $\|\cdot\|_\infty$ given by

$$\|f\|_\infty = \sup_{x \in [a,b]} |f(x)|$$

is well-defined for any $f \in C[a, b]$, and it can be shown that $\|\cdot\|_\infty$ is a norm for $C[a, b]$. Alternatively, since any function f in this vector space is continuous, the function $|f|$ is Riemann integrable, and thus

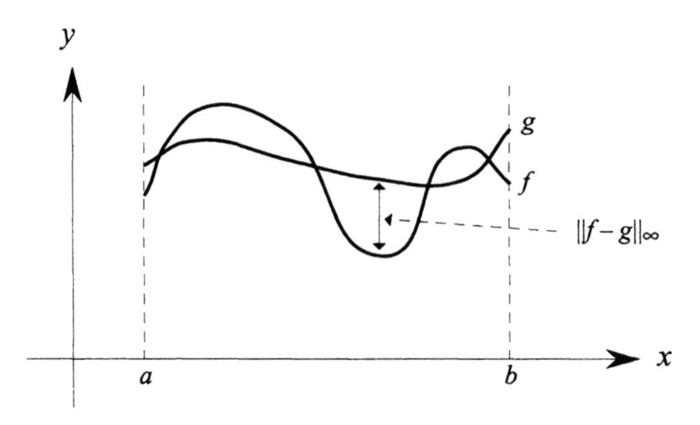

FIGURE 8.1

the function $\|\cdot\|_R$ given by

$$\|f\|_R = \int_a^b |f(x)|\,dx$$

is well-defined on $C[a, b]$. It is left as an exercise to show that $\|\cdot\|_R$ is a norm on $C[a, b]$.

The above examples indicate that a given vector space may have several norms leading to different normed vector spaces. For this reason, the notation $(X, \|\cdot\|)$ is often used to denote the vector space X equipped with the norm $\|\cdot\|$.

Once a vector space is equipped with a norm $\|\cdot\|$, a generalized distance function (called the metric induced by the norm $\|\cdot\|$) can be readily defined. The distance $d(f, g)$ of an element $f \in X$ from another element $g \in X$ is defined to be

$$d(f, g) = \|f - g\|.$$

The distance function for the normed vector space $(\mathbb{R}^n, \|\cdot\|_e)$ corresponds to the ordinary notion of Euclidean distance. The distance function for the normed vector space $(C[a, b], \|\cdot\|_\infty)$ measures the maximum vertical separation of the graph of f from the graph of g (Figure 8.1).

Convergence can be defined for sequences in a normed vector space in a manner that mimics the familiar definition in real analysis. Let $(X, \|\cdot\|)$ be a normed vector space and let $\{f_n\}$ denote an infinite sequence in X. The sequence $\{f_n\}$ is said to **converge** in the norm if there exists an $f \in X$ such that for every $\epsilon > 0$ an integer N can be found with the property that $\|f_n - f\| < \epsilon$ whenever $n > N$. The element f is called the **limit** of the sequence $\{f_n\}$, and the relationship is denoted by $\lim_{n\to\infty} f_n = f$ or simply $f_n \to f$. Note that convergence depends on the choice of norm: A sequence may converge in one norm and diverge in another. Note also that the limit f must also be an element in X.

In a similar spirit, we can define Cauchy sequences for a normed vector space. A sequence $\{f_n\}$ in X is a **Cauchy sequence** (in the norm $\|\cdot\|$) if for any $\epsilon > 0$ there is an integer N such that

$$\|f_m - f_n\| < \epsilon$$

whenever $m > N$ and $n > N$. Cauchy sequences play a vital role in the theory of normed vector spaces. As with convergence, a sequence $\{f_n\}$ in X may be a Cauchy sequence for one choice of norm but not a Cauchy sequence for another choice.

It may be possible to define any number of norms on a given vector space X. Two different norms, however, may yield exactly the same results concerning convergence and Cauchy sequences. Two norms $\|\cdot\|_a$ and $\|\cdot\|_b$ on a vector space X are said to be **equivalent** if there exists positive numbers α and β such that for all $f \in X$,

$$\alpha\|f\|_a \le \|f\|_b \le \beta\|f\|_a.$$

If the norms $\|\cdot\|_a$ and $\|\cdot\|_b$ are equivalent, then it is straightforward to show that convergence in one norm implies convergence in the other, and that the set of Cauchy sequences in $(X, \|\cdot\|_a)$ is the same as the set of Cauchy sequences in $(X, \|\cdot\|_b)$. Equivalent norms lead to the same analytical results.

Identifying norms as equivalent can be an involved process. In finite-dimensional vector spaces, however, the situation is simple: All norms defined on a finite-dimensional vector space are equivalent. Thus the two norms defined in Example 8-1-4 are equivalent. The situation is different for infinite-dimensional spaces. For example, the norms $\|\cdot\|_R$ and $\|\cdot\|_\infty$ defined on the space $C[a, b]$ in Example 8-1-5 are not equivalent. We elucidate further this comment in the next section, when we discuss completeness.

Exercises 8-1:

1. Let $\mathbb{Q}$ denote the set of rational numbers. Show that $\mathbb{Q}$ is a vector space, provided that the scalar field is the rational numbers.

2. (a) Prove that the function $\|\cdot\|_R$ defined on $C[a,b]$ in Example 8-1-5 satisfies the conditions of a norm.

   (b) Suppose that the set $C[a,b]$ is extended to $R[a,b]$, the set of all functions $f : [a,b] \to \mathbb{R}$ such that $|f|$ is Riemann integrable. Show that $\|\cdot\|_R$ is not a norm on $R[a,b]$.

3. Let $C^n[a,b]$ denote the set of functions $f : [a,b] \to \mathbb{R}$ with at least n continuous derivatives on $[a,b]$. Show that the functions $\|\cdot\|_{1,\infty}$ and $\|\cdot\|_{1,1}$ defined by

$$\|f\|_{1,\infty} = \sup_{x\in[a,b]} |f(x)| + \sup_{x\in[a,b]} |f'(x)|,$$

$$\|f\|_{1,1} = \int_{[a,b]} \left\{|f(x)| + |f'(x)|\right\} dx,$$

   are norms on the space $C^1[a,b]$.

4. The number $\sqrt{2}$ can be approximated by a sequence $\{a_n\}$ of rational numbers. Let $S_0 = \{1, 2, \ldots, 9\}$ and choose a_0 as the largest element of S_0 such that $a_0^2 < 2$. Since $1^2 = 1 < 2^2 = 4$, we have $a_0 = 1$. Let $S_1 = \{1.1, 1.2, \ldots, 1.9\}$ and choose a_1 as the largest element of S_1 such that $a_1^2 < 2$. Thus, $a_1 = 1.4$. Let $S_2 = \{1.41, 1.42, \ldots, 1.49\}$ and choose a_2 as the largest element in S_2 such that $a_2^2 < 2$. This gives $a_2 = 1.41$. Following this procedure for the general n show that the resulting sequence $\{a_n\}$ must be a Cauchy sequence.

5. Suppose that $\|\cdot\|_a$ and $\|\cdot\|_b$ are equivalent norms for the vector space X. Prove that the condition

$$\alpha\|f\|_a \le \|f\|_b \le \beta\|f\|_a,$$

   where α and β are positive numbers, implies that there exist positive numbers γ and δ such that

$$\gamma\|f\|_b \le \|f\|_a \le \delta\|f\|_b.$$

8.2 Banach Spaces

The definitions for convergence and Cauchy sequences for the normed vector space $(\mathbb{R}^n, \|\cdot\|_e)$ correspond to the familiar definitions given in real analysis. Various results such as the uniqueness of the limit can be proved in general normed vector spaces by essentially the same techniques used to prove analogous results in real analysis. The space $(\mathbb{R}^n, \|\cdot\|_e)$, however, has a special property not inherent in the definition of a normed vector space. It is well known that a sequence in $(\mathbb{R}, \|\cdot\|_e)$ converges if and only if it is a Cauchy sequence. This result does not extend to the general normed vector space. It is left as an exercise to show that every convergent sequence in a normed vector space must be a Cauchy sequence. The converse is not true. The following examples illustrate the problem for finite- and infinite-dimensional spaces.

Example 8-2-1:
The set $\mathbb{Q}$ of rational numbers, equipped with the Euclidean norm $\|\cdot\|_{\hat{e}}$ restricted to the rational numbers, is a normed vector space, provided that the scalar field is the rational numbers (Exercises 8-1, No. 1). It is well known that the number $\sqrt{2}$ is not a rational number. The sequence $\{a_n\}$ defined in Exercises 8-1, No. 4, is a Cauchy sequence, which in the normed vector space $(\mathbb{R}, \|\cdot\|_e)$ can be shown to converge to the limit $\sqrt{2}$. This sequence is also a Cauchy sequence in $\mathbb{Q}$, but it cannot converge to an element in $\mathbb{Q}$ and is therefore not convergent in $\mathbb{Q}$.

Example 8-2-2:
Consider the normed vector space $(C[-1,1], \|\cdot\|_R)$ and the sequence of functions $\{f_n\}$ defined by

$$f_n(x) = \begin{cases} 1, & \text{if } -1 \leq x \leq 0, \\ 1 - 2^n x, & \text{if } 0 < x \leq 1/2^n, \\ 0, & \text{if } 1/2^n < x \leq 1. \end{cases}$$

The function f_n is depicted in Figure 8.2, and it is clear that $f_n \in C[-1,1]$ for all positive integers n.

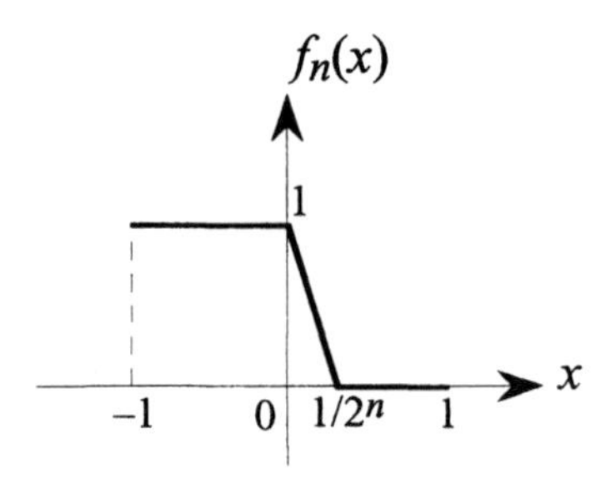

FIGURE 8.2

Now, $|f_n| = f_n$, and

$$\|f_n\|_R = \int_{-1}^{1} f_n(x)\, dx = 1 + \frac{1}{2^{n+1}};$$

therefore, for any $m > N, n > N$, we have that

$$\|f_m - f_n\|_R = \left| \frac{1}{2^{m+1}} - \frac{1}{2^{n+1}} \right| < \frac{1}{2^N}.$$

Given any $\epsilon > 0$, any positive integer N such that $N > -\log_2 \epsilon$ suffices to ensure that $\|f_m - f_n\|_R < \epsilon$ whenever $n > N$ and $m > N$. The sequence is thus a Cauchy sequence.

It is clear geometrically that f_n approaches the function f defined by

$$f(x) = \begin{cases} 1, & \text{if } -1 \le x \le 0, \\ 0, & \text{if } 0 < x \le 1. \end{cases}$$

Indeed, for any *fixed* $x_0 \in [-1, 1]$ the sequence of real numbers $\{f_n(x_0)\}$ converges to $f(x_0)$ (in the $\|\cdot\|_e$ norm), i.e., $\{f_n\}$ is pointwise convergent to f. The function f, however, cannot be a limit in $C[-1, 1]$ for $\{f_n\}$ because $f \notin C[-1, 1]$.

Although the pointwise limit function f cannot be a limit in $C[-1, 1]$ for $\{f_n\}$, this does not extinguish the possibility that there is some other function $g \in C[-1, 1]$ that is the limit. We will show now that no such g exists. Suppose, for a contradiction, that $f_n \to g \in C[-1, 1]$ in the $\|\cdot\|_R$ norm. Then

$$I = \int_{-1}^{1} |f(x) - g(x)|\, dx = \int_{-1}^{1} |(f(x) - f_n(x)) + (f_n(x) - g(x))|\, dx$$

$$
\begin{aligned}
&\le \int_{-1}^{1} |f_n(x) - f(x)|\,dx + \int_{-1}^{1} |f_n(x) - g(x)|\,dx \\
&= \frac{1}{2^{n+1}} + \int_{-1}^{1} |f_n(x) - g(x)|\,dx \\
&= I_f + I_g.
\end{aligned}
$$

The quantity I_f can be made arbitrarily small by choosing n sufficiently large. By hypothesis $f_n \to g$ in the $\|\cdot\|_R$ norm, so that I_g can be made arbitrarily small for n large. Now, $I \le I_f + I_g$, and I does not depend on n. This implies that $I = 0$. Since $g \in C[-1,1]$ and f is continuous on the intervals $[-1,0), (0,1]$, the condition $I = 0$ implies that $f = g$ for all $x \in [-1,0)$ and $x \in (0,1]$. Therefore, $\lim_{x\to 0^-} g(x) = 1$ and $\lim_{x\to 0^+} g(x) = 0$, so that $\lim_{x\to 0} g(x)$ does not exist, contradicting the assumption that g is continuous on the interval $[-1,1]$. The Cauchy sequence $\{f_n\}$ therefore does not converge in the $\|\cdot\|_R$ norm.

Suppose that the space $C[-1,1]$ is equipped with the $\|\cdot\|_\infty$ norm defined in Example 8-1-5 instead of the $\|\cdot\|_R$ norm. If $\{h_n\}$ is a Cauchy sequence in the $\|\cdot\|_\infty$ norm, then $\{h_n\}$ converges pointwise to some limit function h. The difference here is that because $\{h_n\}$ is a Cauchy sequence in the $\|\cdot\|_\infty$ norm, it can be shown that $\{h_n\}$ in fact converges *uniformly* to h, and a standard result in real analysis implies that a uniformly convergent sequence of continuous functions converges to a continuous function. In other words, the limit function h must be in $C[-1,1]$. It is left as an exercise to verify that the sequence $\{f_n\}$ defined in this example is not a Cauchy sequence in the $\|\cdot\|_\infty$ norm.

A normed vector space is called **complete** if every Cauchy sequence in the vector space converges. Complete normed vector spaces are called **Banach spaces**. In finite-dimensional vector spaces, completeness in one norm implies completeness in any norm, since all norms are equivalent. Thus, spaces such as $(\mathbb{R}^n, \|\cdot\|_e)$ and $(\mathbb{R}^n, \|\cdot\|_T)$ are Banach spaces. On the other hand, Example 8-2-1 shows that no norm on the vector space $\mathbb{Q}$ can be defined so that the resulting space is a Banach space. For finite-dimensional vector spaces, completeness depends entirely on the vector space; for infinite-dimensional vector spaces Example 8-2-2 shows that completeness depends also on the choice of norm. The space $(C[-1, 1], \|\cdot\|_\infty)$ is a Banach space, whereas, the space $(C[-1, 1], \|\cdot\|_R)$ is not. If the norms $\|\cdot\|_a$ and $\|\cdot\|_b$ are equivalent, then the corresponding normed vector spaces are either both Banach or both incomplete, since the set of Cauchy sequences is the same for each space and convergence in one norm implies convergence in the other. The two norms $\|\cdot\|_R$ and $\|\cdot\|_\infty$ on $C[-1, 1]$ are evidently not equivalent.

Given a set $S \subseteq X$, if $(X, \|\cdot\|)$ is a normed vector space, a new subset $\overline{S}$ called the **closure** can be formed by requiring that $f \in \overline{S}$ if and only if there is a sequence $\{f_n\}$ of vectors in S (not necessarily distinct) such that $f_n \to f$ (in the norm $\|\cdot\|$). If $S = \overline{S}$, then S is called a **closed** set. The subset S is called **complete** if every Cauchy sequence in S converges to a vector in S. For Banach spaces the concepts of completeness and closure are linked by the following fundamental result:

Theorem 8.2.1
Let $(X, \|\cdot\|)$ be a Banach space and $S \subseteq X$. The set S is closed if and only if S is complete.
In particular, if S is a subspace, then it forms a normed vector space $(S, \|\cdot\|)$ in its own right, and if it is closed, the above theorem indicates that $(S, \|\cdot\|)$ is a Banach space. This observation leads to the following corollary:

Corollary 8.2.2
Let $(X, \|\cdot\|)$ be a Banach space and $S \subseteq X$. Then $(\overline{[S]}, \|\cdot\|)$ is also a Banach space.

Exercises 8-2:

1. Let $(X, \|\cdot\|)$ be a normed vector space and suppose that $\{a_n\}$ is a sequence in X that converges in the norm to some element $a \in X$. Prove that $\{a_n\}$ must be a Cauchy sequence.

2. Let $\{f_n\}$ be the sequence defined in Example 8-2-2. Prove that $\{f_n\}$ is not a Cauchy sequence in the normed vector space $(C[-1, 1], \|\cdot\|_\infty)$.

8.3 Completion of Spaces

If a normed vector space is not complete, it is possible to "expand" the vector space and suitably redefine the norm so that the resulting space is complete. In this section we discuss this process, and in the next section we apply the result to get a Banach space with a norm defined by the Lebesgue integral. Before we discuss the main result, however, we need to introduce a few terms.

In functional analysis, a function $T : X \to Y$ that maps a normed vector space X to a normed vector space Y is called an **operator**.

Example 8-3-1:
Let the set $C^1[a, b]$ and the norm $\|\cdot\|_{1,\infty}$ be as defined in Exercises 8-1, No. 3. Every function in $C^1[a, b]$ has a continuous derivative. If T is the operator corresponding to differentiation d/dx, then T maps every element in $C^1[a, b]$ to a unique function continuous on the interval $[a, b]$. Thus T maps $C^1[a, b]$ into $C[a, b]$. The definition of an operator is not norm dependent, but for illustration, we can regard T as mapping the space $(C^1[a, b], \|\cdot\|_{1,\infty})$ into the space $(C[a, b], \|\cdot\|_\infty)$.

Much of functional analysis is concerned with the study of operators. For a general discussion the reader can consult [25]. Here, we limit ourselves to a special type of operator that preserves norm. An operator T from the normed vector space $(X, \|\cdot\|_X)$ into the normed vector space $(Y, \|\cdot\|_Y)$ is called an **isometry** if for all $x_1, x_2 \in X$

$$\|x_1 - x_2\|_X = \|Tx_1 - Tx_2\|_Y.$$

In essence, an isometry preserves the distance between points in X when they are mapped to Y. An isometry must be one-to-one. If there exists an isometry $T : X \to Y$ such that T is onto (so the inverse $T^{-1} : Y \to X$ exists), then the normed spaces $(X, \|\cdot\|_X)$ and $(Y, \|\cdot\|_Y)$ are called **isometric**. If two spaces are isometric, then completeness in one space implies completeness in the other.

Example 8-3-2:
The differentiation operator of Example 8-3-1 is clearly not an isometry from $(C^1[a, b], \|\cdot\|_{1,\infty})$ into the space $(C[a, b], \|\cdot\|_\infty)$, since in

general,

$$\begin{aligned}\|f-g\|_{1,\infty} &= \sup_{x\in[a,b]} |f(x)-g(x)| + \sup_{x\in[a,b]} |f'(x)-g'(x)| \\ &> \sup_{x\in[a,b]} |f'(x)-g'(x)| = \|f'-g'\|_{\infty}.\end{aligned}$$

Example 8-3-3:
Let $H(\bar{D}(c;r))$ denote the set of all functions holomorphic (analytic) in the closed disk $\bar{D}(c;r) = \{z \in \mathbb{C} : |z-c| \leq r\}$, $r > 0$. This set forms a complex vector space, and the function $\|\cdot\|_c$ defined by

$$\|f\|_c = \sup_{z\in\bar{D}(c;r)} |f(z)|$$

provides a norm for the space. In fact, it can be shown that $(H(\bar{D}(c;r)), \|\cdot\|_c)$ is a Banach space. Let $T_{\phi,b}$ be the operator mapping $H(\bar{D}(a;r))$ to $H(\bar{D}(a-b;r))$ defined by

$$T_{\phi,b}f = e^{i\phi}f(z-b),$$

where $\phi \in \mathbb{R}$ is a constant. The operator $T_{\phi,b}$ is a one-to-one and onto mapping from $(H(\bar{D}(a;r)), \|\cdot\|_a)$ to $(H(\bar{D}(a-b;r)), \|\cdot\|_{a-b})$, and since

$$\begin{aligned}\|T_{\phi,b}f - T_{\phi,b}g\|_{a-b} &= \sup_{z\in\bar{D}(a-b;r)} |e^{i\phi}f(z-b) - e^{i\phi}g(z-b)| \\ &= \sup_{z\in\bar{D}(a-b;r)} |f(z-b)-g(z-b)| \\ &= \sup_{z\in\bar{D}(a;r)} |f(z)-g(z)| = \|f-g\|_a,\end{aligned}$$

the operator is also an isometry. The two normed spaces are thus isometric.

Given an incomplete normed vector space $(X, \|\cdot\|)$, it is natural to enquire whether the space can be made complete by enlarging the vector space and extending the definition of the norm to cope with the new elements. The paradigm for this process is the completion of the rational number system $\mathbb{Q}$ to form the real number system $\mathbb{R}$. This example has two features, which, loosely speaking, are as follows:

(i) the completion does not change the value of the norm where it was originally defined, i.e., $\|r\|_{\hat{e}} = \|r\|_e$, for any rational number r;

(ii) the set $\mathbb{Q}$ is dense in the set $\mathbb{R}$ (cf. Section 1-1).

The first feature is obviously desirable: We wish to preserve as much as possible the original normed vector space, and any extended definition of the norm should not change the value of the norm at points in the original space. The second feature expresses the fact that the extension of the set $\mathbb{Q}$ to the set $\mathbb{R}$ is a "minimal" one: Every element added to $\mathbb{Q}$ is required for the completion. We could have "completed" $\mathbb{Q}$ by including all the complex numbers to form the complex plane $\mathbb{C}$, which is complete, but this is overkill.

The completion of the rational numbers serves as a model for the general completion process. Feature (i) can be framed for general normed vector spaces in terms of isometries. In order to discuss feature (ii) in a general context, however, we need to introduce a general definition of density. Let $(X, \|\cdot\|)$ be a normed vector space and $W \subseteq X$. The set W is **dense** in X if every element of X is the limit of some sequence in W. Density is an important property from a practical viewpoint. If W is dense in X, then any element in X can be approximated by a sequence in W to any degree of accuracy. For example, the sequence $\{a_n\}$ of Example 8-2-1 consists purely of rational numbers and can be used to approximate $\sqrt{2}$ to within any given (nonzero) error.

A fundamental result in the theory of normed vector spaces is that any normed vector space can be completed. Specifically, we have the following result:

Theorem 8.3.1

Given a normed vector space $(X, \|\cdot\|_X)$, there exists a Banach space $(Y, \|\cdot\|_Y)$ containing a subspace $(W, \|\cdot\|_Y)$ with the following properties:

(i) $(W, \|\cdot\|_Y)$ is isometric with $(X, \|\cdot\|_X)$;

(ii) W is dense in Y.

The space $(Y, \|\cdot\|_Y)$ is unique except for isometries. In other words if $(\hat{Y}, \|\cdot\|_{\hat{Y}})$ is also a Banach space with a subspace $(\hat{W}, \|\cdot\|_{\hat{Y}})$ having properties (i) *and* (ii)*, then $(\hat{Y}, \|\cdot\|_{\hat{Y}})$ is isometric with $(Y, \|\cdot\|_Y)$.*

The space $(Y, \|\cdot\|_Y)$ is called the **completion** of the space $(X, \|\cdot\|_X)$. The proof of this result would lead us too far astray from our main subject, integration. We refer the reader to [25] for the details.

Exercises 8-3:

1. (a) Suppose that Z is dense in W, and W is dense in Y. Prove that Z is dense in Y.

   (b) Suppose that the completion of $(X, \|\cdot\|_X)$ is $(Y, \|\cdot\|_Y)$ and that P is dense in X. Prove that $(Y, \|\cdot\|_Y)$ is also the completion of $(P, \|\cdot\|_X)$.
2. Let $P[a, b]$ denote the set of polynomials on the interval $[a, b]$, and let $P_{\mathbb{Q}}[a, b]$ denote the set of polynomials on $[a, b]$ with rational coefficients. Prove that $P_{\mathbb{Q}}[a, b]$ is dense in $P[a, b]$.
3. *Weierstrass*'s theorem asserts that any function in $C[a, b]$ can be approximated uniformly by a sequence of polynomials, i.e., $P[a, b]$ is dense in $C[a, b]$ with repect to the $\|\cdot\|_\infty$ norm. Use Exercises 8-3, No. 2, to deduce that any function in $C[a, b]$ can be approximated uniformly by a sequence in $P_{\mathbb{Q}}[a, b]$.

8.4 The Space L^1

Having made our brief foray into functional analysis, we are now ready to return to matters directly involved with integration. Example 8-2-2 shows that the normed vector space $(C[a, b], \|\cdot\|_R)$ is not complete. We *know*, however, that this space can be completed, but it is not clear exactly what kinds of functions are required to complete it. In this regard, the norm itself can be used as a rough guide. Clearly, a function f need not be in $C[a, b]$ for the Riemann integral of $|f|$ to be defined. This observation suggests that perhaps the appropriate vector space would be $R[a, b]$, the set of all functions $f : [a, b] \to \mathbb{R}$ such that $|f|$ is Riemann integrable. This "expansion" of $C[a, b]$ to $R[a, b]$ solves the immediate problem, since the sequence $\{f_n\}$ in Example 8-2-2 would converge to a function $f \in R[a, b]$, but it opens the floodgates to sequences such as that defined in Section

3-3-1 that do not converge to functions in $R[a,b]$. Although $\|\cdot\|_R$ is not a norm on $R[a,b]$ (Exercises 8-2, No. 2(b)), this problem can be overcome. Any hopes of using $R[a,b]$ to complete the space, however, are dashed by Example 3-3-1, because this example indicates that $(R[a,b], \|\cdot\|_R)$ is not complete.

Recall that Example 4-3-1 motivated us initially to seek a more general integral to accommodate functions such as

$$f(x) = \begin{cases} 1, & \text{if } x \text{ is rational, } x \neq 0, 1, \\ 0, & \text{if } x \text{ is irrational or } x = 0, 1. \end{cases}$$

Eventually, we arrived at the Lebesgue integral. The function f plays a role in the completion of $(R[a,b], \|\cdot\|_R)$ analogous to that played by the number $\sqrt{2}$ in the completion of $(\mathbb{Q}, \|\cdot\|_{\hat{e}})$. The Lebesgue integral essentially leads us to the appropriate space and isometry for the completion of $(R[a,b], \|\cdot\|_R)$ (and $(C[a,b], \|\cdot\|_R)$).

Let $\Lambda^1[a,b]$ denote the space of all functions $f : [a,b] \to \mathbb{R}$ that are (Lebesgue) integrable on the interval $[a,b]$ and let $\|\cdot\|_1$ be the function defined by

$$\|f\|_1 = \int_{[a,b]} |f(x)|\, dx.$$

The set $\Lambda^1[a,b]$ forms a vector space, but $\|\cdot\|_1$ is not a norm on it because there are nonzero functions g in $\Lambda^1[a,b]$ such that $\|g\|_1 = 0$, i.e., if $g = 0$ a.e. then $\|g\|_1 = 0$. Functions that fail to be norms solely because they cannot satisfy this condition are called **seminorms**, and the resulting spaces are called **seminormed vector spaces**. Notions such as convergence and Cauchy sequences are defined for seminormed vector spaces in the same way they are defined for normed vector spaces.

The problem with the seminorm on $\Lambda^1[a,b]$ is not insurmountable. The essence of the problem is that $\|g\| = \|f\|$ whenever $f = g$ a.e. (Theorem 5.2.3 (iii)). The set $\Lambda^1[a,b]$, however, can be partitioned into equivalence classes based on equality a.e. Let $L^1[a,b]$ denote the set of equivalence classes of $\Lambda^1[a,b]$. An element F of $L^1[a,b]$ is thus a set of functions such that if $f_1, f_2 \in F$, then $f_1 = f_2$ a.e. Since any element f of F can be used to represent the equivalence class, we use the notation $F = [f]$.[1] Addition is defined as

$$[f] + [g] = [f + g],$$

and scalar multiplication as

$$\alpha[f] = [\alpha f].$$

The set $L^1[a, b]$ forms a vector space, and if $\| \cdot \|_1$ is defined by

$$\|[f]\| = \int_{[a,b]} |f(x)|\, dx,$$

then $(L^1[a, b], \| \cdot \|_1)$ is a normed vector space.

The candidate for the completion of the space $(C[a, b], \| \cdot \|_R)$ (and the space $(R[a, b], \| \cdot \|_R)$) is the space $(L^1[a, b], \| \cdot \|_1)$. In the notation of the previous section, we have $X = C[a, b]$, $\| \cdot \|_X = \| \cdot \|_R$, $Y = L^1[a, b]$, and $\| \cdot \|_Y = \| \cdot \|_1$. Let $W = \{[f] \in L^1[a, b] : [f]$ contains a function in $C[a, b]\}$, and let T be the operator that maps a function $f \in C[a, b]$ to the element $[f] \in W$. Now, every element of $C[a, b]$ has a corresponding element in W, and no equivalence class in W contains two distinct functions in $C[a, b]$; therefore, T is a one-to-one, onto operator from $C[a, b]$ to W. Moreover, Theorem 4.6.1 implies that

$$\begin{aligned} \|Tf\|_1 = \|[f]\|_1 &= \int_{[a,b]} |f(x)|\, dx \\ &= \int_a^b |f(x)|\, dx = \|f\|_R, \end{aligned}$$

so that T is an isometry. The space W is thus isometric with $C[a, b]$. To establish that $(L^1[a, b], \| \cdot \|_1)$ is the completion of $(C[a, b], \| \cdot \|_R)$ it remains to show that W is dense in $L^1[a, b]$ and that $(L^1[a, b], \| \cdot \|_1)$ is a Banach space. We will not prove that W is dense in $L^1[a, b]$. The reader is referred to [37] for this result. We will, however, sketch a proof that $(L^1[a, b], \| \cdot \|_1)$ is complete.

Theorem 8.4.1
The normed vector space $(L^1[a, b], \| \cdot \|_1)$ is a Banach space.

---

[1] Although this is standard notation, there is some danger of confusion with the notation used for the *span* that takes sets as arguments

Proof We prove that the seminormed space $(\Lambda^1[a,b], \|\cdot\|_1)$ is complete. The completeness of $(L^1[a,b], \|\cdot\|_1)$ then follows upon identification of the functions with their equivalence classes in $L^1[a,b]$. Let $\{f_n\}$ be a Cauchy sequence in $(\Lambda^1[a,b], \|\cdot\|_1)$. Given any $\epsilon > 0$ there is thus an integer N such that $\|f_n - f_m\| < \epsilon$ whenever $m > N$ and $n > N$. In particular, there is a subsequence $\{f_{n_k}\}$ of $\{f_n\}$ with the property that

$$\|f_{n_{k+1}} - f_{n_k}\|_1 < \frac{1}{2^k}.$$

Let

$$g_m = \sum_{k=1}^{m} |f_{n_{k+1}} - f_{n_k}|,$$

and let $g = \lim_{m\to\infty} g_m$ denote the pointwise limit function. Note that $g(x)$ need not be finite for all $x \in [a,b]$. Let $[a,b] = I_1 \cup I_2$, where I_1 denotes the set of all points such that $g(x) < \infty$. We will show that I_2 must be a null set.

Now, $g_m \in \Lambda^1[a,b]$ and

$$\begin{aligned}\|g_m\|_1 &= \int_{[a,b]} |g_m(x)|\,dx \leq \sum_{k=1}^{m} \int_{[a,b]} |f_{n_{k+1}}(x) - f_{n_k}(x)|\,dx \\ &= \sum_{k=1}^{m} \|f_{n_{k+1}} - f_{n_k}\|_1 < \sum_{k=1}^{m} \frac{1}{2^k} < 1.\end{aligned}$$

The sequence $\{g_m\}$ is a monotone sequence of functions in $\Lambda^1[a,b]$, and the above inequalities indicate that $\lim_{m\to\infty} \|g_m\|_1 \leq 1$. The monotone convergence theorem (Theorem 5.3.1) implies that $g \in \Lambda^1[a,b]$ and $\|g_m\|_1 \to \|g\|_1$; hence, $\|g\|_1 \leq 1$. Since $\|g\|_1$ is finite, $g(x) < \infty$ a.e., and so the set I_2 must be null. The series

$$f_{n_1}(x) + \sum_{k=1}^{\infty} (f_{n_{k+1}}(x) - f_{n_k}(x))$$

must therefore be absolutely convergent for almost all x. This series thus defines a function f, the pointwise limit, almost everywhere. Eventually, f will be identified with an equivalence class in $L^1[a,b]$, so the fact that f is defined only a.e. is not a real concern.

We have shown that $f_{n_k} \to f$; we need to show that $f_n \to f$ in the $\|\cdot\|_1$ seminorm and that $f \in \Lambda^1[a,b]$. Since $\{f_n\}$ is a Cauchy sequence, for any $\epsilon > 0$ there is an integer N such that

$$\|f_n - f_m\|_1 = \int_{[a,b]} |f_n(x) - f_m(x)|\, dx < \epsilon$$

for any $m > N, n > N$. Let k be sufficiently large so that $n_k > N$ and let $m = n_k$. Then for $n > N$,

$$0 \leq \lim_{k\to\infty} \left(\inf_{m \geq k} \int_{[a,b]} |f_n(x) - f_{n_m}(x)|\, dx \right) \leq \epsilon,$$

and so Fatou's lemma (Lemma 5.3.2) implies that for $n > N$, $|f_n - f|$ is integrable over $[a,b]$ and $\int_{[a,b]} |f_n(x) - f(x)|\, dx \leq \epsilon$. Therefore, $f_n - f \in \Lambda^1[a,b]$, and so $f \in \Lambda^1[a,b]$. Moreover, $\|f_n - f\|_1 \to 0$ as $n \to \infty$, so that the sequence $\{f_n\}$ converges to f in the $\|\cdot\|_1$ seminorm. The space $(\Lambda^1[a,b], \|\cdot\|_1)$ is thus complete. The completeness of this space implies the completeness of the space $(L^1[a,b], \|\cdot\|_1)$, since each Cauchy sequence in $L^1[a,b]$ can be represented by a Cauchy sequence in $\Lambda^1[a,b]$. □

8.5 The Lebesgue Spaces L^p

The norm defined for the space $L^1[a,b]$ is a "natural" choice in applications where the average magnitude of a function is of conspicuous importance. The function

$$\|f\|_1 = \int_{[a,b]} |f(x)|\, dx$$

is the continuous analogue of the norm $\|\cdot\|_T$ defined in Example 8-1-4 for $\mathbb{R}^n$. If we seek a continuous analogue for the Euclidean norm in $\mathbb{R}^n$ we are led to the function $\|\cdot\|_2$ defined by

$$\|f\|_2 = \left\{ \int_{[a,b]} |f(x)|^2\, dx \right\}^{1/2},$$

and more generally, if we seek a continuous analogue to a general mean norm for $\mathbb{R}^n$,

$$\|\mathbf{x}\| = \left\{|x_1|^p + |x_2|^p + \cdots + |x_n|^p\right\}^{1/p}$$

for $p \geq 1$, we are led to a function $\|\cdot\|_p$ defined by

$$\|f\|_p = \left\{\int_{[a,b]} |f(x)|^p\,dx\right\}^{1/p}.$$

In this manner, vector spaces for which these functions define norms or seminorms come into prominence.

The space $(L^1[a,b], \|\cdot\|_1)$ serves as a prototype for all the Lebesgue spaces. Let $\Lambda^p[a,b]$, $1 \leq p < \infty$, denote the set of measurable functions f such that

$$\int_{[a,b]} |f(x)|^p\,dx < \infty.$$

Now, $\|f\|_p = \|g\|_p$ for any $f, g \in \Lambda^p[a,b]$ such that $f = g$ a.e., so we know that $\|\cdot\|_p$ is at best a seminorm for $\Lambda^p[a,b]$. This problem can be easily remedied by using equivalence classes. A more serious concern is that $\Lambda^p[a,b]$ may not even be a vector space. In particular, if $f, g \in \Lambda^p[a,b]$, it is not clear that $f + g \in \Lambda^p[a,b]$. Moreover, it is not obvious that $\|\cdot\|_p$ will satisfy the triangle inequality. As it turns out, the sets $\Lambda^p[a,b]$ are vector spaces and $\|\cdot\|_p$ is a seminorm on them for $1 \leq p < \infty$. This follows from Minkowski's inequality, which is derived from another inequality of importance called Hölder's inequality (versions of these results are given below for the corresponding L^p spaces).

Let $L^p[a,b]$ denote the set of the equivalence classes of $\Lambda^p[a,b]$ modulo equivalence a.e., and for $[f] \in L^p[a,b]$ define the function $\|\cdot\|_p$ by

$$\|[f]\|_p = \left\{\int_{[a,b]} |f(x)|^p\,dx\right\}^{1/p}.$$

Theorem 8.5.1 (Hölder's Inequality)
Let $F \in L^p[a,b]$ and $G \in L^q[a,b]$, where $1 < p < \infty$ and $1/p + 1/q = 1$. Then $FG \in L^1[a,b]$ and

$$\|FG\|_1 \leq \|F\|_p \|G\|_q.$$

Theorem 8.5.2 (Minkowski's Inequality)
Let $1 < p < \infty$ and suppose that $F, G \in L^p[a,b]$. Then $F+G \in L^p[a,b]$ and

$$\|F+G\|_p \leq \|F\|_p + \|G\|_p.$$

The proofs of these inequalities can be found in most texts on functional analysis, e.g., [38]. In the Hölder inequality, the product is the pointwise product of functions, i.e., if $F = [f]$, $G = [g]$, then $FG = [fg]$, where $(fg)(x) = f(x)g(x)$. That $L^p[a,b]$ is a vector space and $\|\cdot\|_p$ defines a norm on it follows from Minkowski's inequality.

As with the space $(L^1[a,b], \|\cdot\|_1)$, the normed vector spaces $(L^p[a,b], \|\cdot\|_p)$ are complete. This result is a generalized version of the classical *Riesz–Fischer theorem*.

Theorem 8.5.3
The normed vector spaces $(L^p[a,b], \|\cdot\|_p)$ are Banach spaces for $1 \leq p < \infty$.
A detailed proof of this result can be found in [17] and [18]. The proof for the case $1 < p < \infty$ is similar to that for the case $p = 1$. Essentially, the civilized behavior of the Lebesgue integral (as manifested in the monotone convergence theorem) is responsible for completeness. The Lebesgue integral thus yields an entire family of Banach spaces.

To simplify notation, we shall refer to the Banach space $(L^p[a,b], \|\cdot\|_p)$ simply as $L^p[a,b]$ unless there is some ambiguity regarding the norm. These Banach spaces are collectively referred to as the **Lebesgue** or **L^p spaces**. We also follow the common (and convenient) practice of blurring the distinction between $\Lambda^p[a,b]$ and $L^p[a,b]$ by treating elements of $L^p[a,b]$ as functions. We trust the reader to make the correct technical interpretation and to remember in this context that "$f = g$" means $f = g$ a.e.

Suppose that $f \in L^2[a,b]$. The constant function $g = 1$ is also in $L^2[a,b]$, and therefore Hölder's inequality implies that the function $f \cdot 1 = f$ is in $L^1[a,b]$. In addition, we have that

$$\|f\|_1 \leq \|f\|_2 \|1\|_2 = (b-a)^{1/2} \|f\|_2.$$

This observation shows that $L^2[a,b] \subseteq L^1[a,b]$. The calculation "works" because $p = q = 2$ in Hölder's inequality and $g = 1$ is integrable on any interval of *finite* length. We can repeat this argument for the general $p > 1$ because $g \in L^q[a,b]$ for any q. Thus, $L^p[a,b] \subseteq L^1[a,b]$ for all $p > 1$. The next result indicates that if $1 \le p_1 \le p_2$, then $L^{p_2}[a,b] \subseteq L^{p_1}[a,b]$.

Theorem 8.5.4
Let $1 \le p_1 \le p_2$ and suppose that $f \in L^{p_2}[a,b]$. Then

$$\|f\|_{p_1} \le \|f\|_{p_2}(b-a)^{1/p_1-1/p_2},$$

and consequently $f \in L^{p_1}[a,b]$.

Proof If $p_1 = p_2$, the result is trivial. Suppose that $1 \le p_1 < p_2$, and let $k = p_2/p_1$, $m = k/(k-1)$. Now,

$$\int_{[a,b]} \left| |f(x)|^{p_1} \right|^k dx = \int_{[a,b]} |f(x)|^{p_2}\, dx < \infty,$$

since $f \in L^{p_2}[a,b]$, and therefore $|f(x)|^{p_1} \in L^k[a,b]$. Hölder's inequality with $p = k$, $q = m$, and $g = 1 \in L^m[a,b]$ yields the inequality

$$\| |f(x)|^{p_1} \cdot 1 \|_1 \le \| |f(x)|^{p_1} \|_k \|1\|_m.$$

But

$$\| |f(x)|^{p_1} \cdot 1 \|_1 = \int_{[a,b]} |f(x)|^{p_1}\, dx = \left(\|f\|_{p_1}\right)^{p_1},$$

$$\| |f(x)|^{p_1} \|_k = \left\{ \int_{[a,b]} \left(|f(x)|^{p_1}\right)^k dx \right\}^{1/k} = \left(\|f\|_{p_2}\right)^{p_2/k} = \left(\|f\|_{p_2}\right)^{p_1},$$

and

$$\|1\|_m = (b-a)^{1/m} = (b-a)^{1-p_1/p_2}.$$

Consequently,

$$\left(\|f\|_{p_1}\right)^{p_1} \le \left(\|f\|_{p_2}\right)^{p_1} (b-a)^{1-p_1/p_2},$$

and the inequality follows. □

Two positive numbers p and q are called **conjugate exponents** if $1/p + 1/q = 1$. Hölder's inequality suggests that the Banach spaces $L^p[a,b]$ and $L^q[a,b]$ are related if p and q are conjugate exponents. In fact, the spaces are intimately related. A **linear functional** is an operator J from a normed vector space $(X, \|\cdot\|)$ to the scalar field ($\mathbb{R}$ or $\mathbb{C}$) of the vector space such that for any $f_1, f_2 \in X$

$$J(f_1 + f_2) = J(f_1) + J(f_2),$$

and for any scalar α

$$J(\alpha f_1) = \alpha J(f_1).$$

The functional J is **bounded** if there is a number $c > 0$ such that $|J(f)| \leq c\|f\|$ for all $f \in X$. The norm of a bounded functional J is defined as the smallest number c such that $|J(f)| \leq c\|f\|$ for all $f \in X$ and denoted by $\|J\|$. Since $\alpha = 0$ is a legitimate choice for a scalar, we see that

$$J(0 \cdot f_1) = J(\mathbf{0}) = 0J(f_1) = 0,$$

i.e., $J(\mathbf{0}) = 0$. The norm of J is thus given by

$$\|J\| = \sup_{\substack{f \in X \\ f \neq 0}} \frac{|J(f)|}{\|f\|}$$

(any choice of $c > 0$ satisfies $|0| \leq c\|\mathbf{0}\|$).

Example 8-5-1:
Let

$$J(f) = \int_{[a,b]} k(x) f(x)\, dx,$$

where $k \in L^2[a, b]$, and let $X = L^2[a, b]$. Now, J is evidently a linear functional, and Hölder's inequality implies that for any $f \in L^2[a, b]$,

$$|J(f)| \leq \int_{[a,b]} |k(x) f(x)|\, dx = \|kf\|_1 \leq \|k\|_2 \|f\|_2;$$

thus,

$$\|J\| \leq \|k\|_2.$$

In fact, $\|J\| = \|k\|_2$. To see this, note first that if $\|k\|_2 = 0$, then $k = 0$ a.e., and therefore $kf = 0$ a.e. for any $f \in L^2[a, b]$; hence, $J(f) = 0$ for all $f \in L^2[a, b]$ and $\|J\| = 0$. If $\|k\|_2 \neq 0$, choose $f = k/\|k\|_2 \in L^2[a, b]$.

Now,

$$|J(f)| = \left| \int_{[a,b]} k(x) \frac{k(x)}{\|k\|_2} \, dx \right|$$
$$= \frac{1}{\|k\|_2} \int_{[a,b]} k^2(x) \, dx = \|k\|_2.$$

Since $\|f\|_2 = 1$ for this choice, we see that $\|J\| = \|k\|_2$.

The above example is typical of bounded linear functionals on the $L^p[a,b]$ spaces for $1 < p < \infty$. The following representation theorem expresses the situation:

Theorem 8.5.5

Let p and q be conjugate exponents with $1 < p < \infty$, and suppose that $J : L^p[a,b] \to \mathbb{R}$ is a bounded linear functional. Then there exists a unique element $g \in L^q[a,b]$ such that

$$J(f) = \int_{[a,b]} f(x)g(x) \, dx,$$

for all $f \in L^p[a,b]$. Moreover,

$$\|J\| = \|g\|_q.$$

The proof of this fundamental result can be found in [37]. In the more general context of functional analysis, the **dual space** of a normed space $(X, \|\cdot\|)$ is defined to be the set of all bounded linear functionals on X. The above theorem indicates that $L^q[a,b]$ is the dual space of $L^p[a,b]$ when p and q are conjugate exponents with $1 < p < \infty$.

Conjugate exponents and dual spaces bring to the fore two exceptional cases. Each Banach space $L^p[a,b]$ is paired with its dual space $L^q[a,b]$, where $1/p + 1/q = 1$, $1 < p < \infty$. The conjugate exponent of $p = 2$ is $q = 2$, and so $L^2[a,b]$ is its own dual space. The space $L^2[a,b]$ is clearly the only member of the $L^p[a,b]$ spaces with this property, and it turns out that $L^2[a,b]$ has several other special properties not shared by the other $L^p[a,b]$ spaces. The space $L^2[a,b]$ is special because it forms what is known as a Hilbert space. We postpone further discussion about the special properties of $L^2[a,b]$ until the next chapter.

The other exceptional case is the space $L^1[a,b]$. There is no conjugate exponent for $p = 1$, yet there are certainly bounded linear functionals defined on $L^1[a,b]$. What then is the dual space of $L^1[a,b]$? If $|k(x)| \leq M < \infty$ for all $x \in [a,b]$, then the functional J defined by

$$J(f) = \int_{[a,b]} k(x) f(x)\, dx$$

is a bounded linear functional, since

$$|J(f)| \leq M \int_{[a,b]} |f(x)|\, dx = M \|f\|_1.$$

In fact, the condition that $|k(x)| \leq M < \infty$ for all $x \in [a,b]$ can be relaxed to $|k(x)| \leq M < \infty$ a.e. in view of the Lebesgue integral. This example serves as a guide to what might be expected as a dual space for $L^1[a,b]$.

Let $f : [a,b] \to \mathbb{R}$ be a measurable function. A number μ is an **essential upper bound** for $|f|$ if $|f(x)| \leq \mu$ a.e. If $|f|$ has an essential upper bound, then it can be shown that there exists a least such bound. This least bound is denoted by ess sup $|f|$. Let $\Lambda^\infty[a,b]$ denote the set of all measurable functions on $[a,b]$ such that ess sup $|f| < \infty$ and define the function $\|\cdot\|_\infty$ by

$$\|f\|_\infty = \text{ess sup}\, |f|.$$

As with the $\Lambda^p[a,b]$ spaces, $\Lambda^\infty[a,b]$ is a vector space and $\|\cdot\|_\infty$ is a seminorm on it. Let $L^\infty[a,b]$ denote the set of equivalence classes of $\Lambda^\infty[a,b]$ modulo equality a.e. Then $(L^\infty[a,b], \|\cdot\|_\infty)$ is a normed vector space. The following theorem summarizes some of the properties of $L^\infty[a,b]$:

Theorem 8.5.6

(i) $(L^\infty[a,b], \|\cdot\|_\infty)$ *is a Banach space.*

(ii) The dual space of $L^1[a,b]$ *is* $L^\infty[a,b]$.

(iii) If $f \in L^1[a,b]$ *and* $g \in L^\infty[a,b]$, *then* $fg \in L^1[a,b]$ *and*

$$\|fg\|_1 \leq \|f\|_1 \|g\|_\infty$$

(an extension of the Hölder inequality).

(iv) If $f \in L^\infty[a,b]$, *then* $f \in L^p[a,b]$ *for all* $1 \leq p < \infty$.

The proofs of parts (i) and (ii) can be found in [33]. The proofs of parts (iii) and (iv) are left as exercises.

If p and q are conjugate exponents with $p > 1$, then the dual space of $L^p[a,b]$ is $L^q[a,b]$ and vice versa. In the language of functional analysis, the L^p spaces are *reflexive* for $p > 1$. In contrast, the dual space of $L^\infty[a,b]$ is not $L^1[a,b]$. The dual space of $L^\infty[a,b]$ turns out to be a space of measures. More details concerning the dual space of $L^\infty[a,b]$ can be found in [34]. The space $L^1[a,b]$ is thus unusual among the $L^p[a,b]$ spaces in that it is not reflexive.

We have thus far been concerned with L^p spaces where the interval of integration is bounded. The definitions for L^p spaces can be extended to include unbounded intervals of integration. If $I \subseteq \mathbb{R}$ is any interval, the space $L^p(I)$ consists of the set of functions $f : I \to \mathbb{R}$ (i.e., equivalence classes) such that

$$\|f\|_p = \left\{ \int_I |f(x)|^p \, dx \right\}^{1/p} < \infty.$$

All the results stated for the $L^p[a,b]$ spaces carry over to the general $L^p(I)$ spaces with the notable exception of the inclusion results. The proof of Theorem 8.5.4 relies crucially on the fact that the interval is bounded. If the interval is not bounded, then results such as $L^2(I) \subseteq L^1(I)$ are not valid. For example, consider the functions f_1 and f_2 defined by

$$f_1(x) = \begin{cases} 1/\sqrt{x}, & \text{if } 0 < x \le 1, \\ 0, & \text{if } x > 1, \end{cases}$$

$$f_2(x) = \begin{cases} 0, & \text{if } 0 < x \le 1, \\ 1/x, & \text{if } x > 1. \end{cases}$$

Now, $f_1 \in L^1(0,\infty)$, but $f_2 \notin L^1(0,\infty)$; alternatively, $f_1 \notin L^2(0,\infty)$ but $f_2 \in L^2(0,\infty)$. Hence $L^2(0,\infty)$ does not contain $L^1(0,\infty)$, and $L^1(0,\infty)$ does not contain $L^2(0,\infty)$.

Exercises 8-5:

1. Suppose that the function $r : [0,1] \to \mathbb{R}$ is continuous and positive on the interval $[a,b]$.

(a) Prove that for $1 \leq p$ the function ${}_r\|\cdot\|_p$ defined by

$$ {}_r\|f\|_p = \left(\int_{[a,b]} r(x)|f(x)|^p \, dx \right)^{1/p} $$

is a norm on $L^p[a,b]$.

(b) Prove that the norm ${}_r\|\cdot\|_p$ is equivalent to the norm $\|\cdot\|_p$.

2. Let $1 \leq p_1 \leq p_2$ and let I be a bounded interval. Prove that $\|f\|_{p_1} \leq 1 + \|f\|_{p_2}$ and from this deduce that $L^{p_2}(I) \subseteq L^{p_1}(I)$.

3. The *Fredholm integral operator* K is defined by

$$ (Kf)(x) = \int_{[0,1]} k(x,\xi) f(\xi) \, d\xi, $$

where $k : [0,1] \times [0,1] \to \mathbb{R}$ is a given function. Suppose that k is bounded in the square $[0,1] \times [0,1]$. Prove that the operator K maps functions in $L^2[0,1]$ to functions in $L^2[0,1]$.

4. Prove parts (iii) and (iv) of Theorem 8.5.6.

5. If $f \in L^\infty[0,1]$, prove that

$$ \|f\|_1 \leq \|f\|_2 \leq \cdots \leq \|f\|_n \leq \|f\|_{n+1} \leq \cdots \leq \|f\|_\infty. $$

8.6 Separable Spaces

A Banach space $(X, \|\cdot\|)$ is **separable** if X contains a *countable* set that is dense in X in the $\|\cdot\|$ norm. The Banach space $(\mathbb{R}, \|\cdot\|_e)$, for example, is a separable space. To establish this claim we note that the rational numbers form a countable set that is dense in the real numbers.

Separability is a desirable property, because the existence of a countable dense set simplifies problems concerning the representation of functions (e.g., bases for spaces) and the approximation of functions. In the Hilbert space theory discussed in Chapter 9, separability is a key ingredient in establishing the existence of an orthonormal basis (cf. Theorem 9.2.4). Fortunately, the L^p spaces are separable.

Theorem 8.6.1
The L^p spaces are separable for $1 \leq p < \infty$.

The proof of this result can be found in [17]. We limit ourselves here to a brief informal discussion of the ideas underlying the proof and focus on sets that are dense in the L^p spaces. These sets are of interest in their own right for the purpose of approximation.

The construction of the Lebesgue–Stieltjes integral in Chapter 4 brings to the fore one class of functions that must be dense in any L^p space, viz. the set of functions consisting of α-summable step functions (Section 4.5). This density relationship is used to prove the Riemann–Lebesgue theorem (Theorem 9.3.4) in the next chapter. Unfortunately, this set is not countable. There are, of course, other sets dense in the L^p spaces. The next result (which we state without proof) provides a sample of such a set.

Theorem 8.6.2
Let $C_0^\infty(\mathbb{R})$ denote the set of functions $f : \mathbb{R} \to \mathbb{R}$ with derivatives of all orders and with supports of finite length. The set $C_0^\infty(\mathbb{R})$ is dense in each $L^p(\mathbb{R})$ space for $1 \leq p < \infty$.
Similar results are available for the $L^p(I)$ spaces where I is an interval.

The set $C_0^\infty(\mathbb{R})$ is not countable, but if the above theorem is coupled with Weierstrass's theorem concerning the approximation of continuous functions by polynomials (Exercises 8-3, No. 3), a countable dense set can be derived. Suppose that $f \in C_0^\infty(\mathbb{R})$ and choose some $\epsilon > 0$. Now, f has a support of finite length, and hence there is some number $\beta > 0$ such that $f = 0$ outside the interval $[-\beta, \beta]$. Certainly f (restricted to the interval $[-\beta, \beta]$) is in the space $C[-\beta, \beta]$, and thus there is a polynomial p_ϵ such that

$$\sup_{x \in [-\beta, \beta]} |p_\epsilon(x) - f(x)| < \frac{\epsilon}{(2\beta)^{1/p}}. \tag{8.1}$$

Moreover, we know from Exercises 8-3-2 that the set $P_{\mathbb{Q}}[-\beta, \beta]$ of polynomials with rational coefficients is dense in $P[-\beta, \beta]$, and hence we can find a polynomial with rational coefficients satisfying inequality (8.1). In fact, we can find a polynomial q_ϵ with rational coefficients such that inequality (8.1) is satisfied and $q_\epsilon(-\beta) = q_\epsilon(\beta) = 0$. Now the polynomial q_ϵ can be extended to a function $g_\epsilon = q_\epsilon \cdot \chi_{[-\beta, \beta]}$

defined on $\mathbb{R}$. Here $\chi_{[-\beta,\beta]}$ is the characteristic function defined by

$$\chi_{[-\beta,\beta]}(x) = \begin{cases} 1, & \text{if } x \in [-\beta, \beta], \\ 0, & \text{otherwise.} \end{cases}$$

Thus,

$$\begin{aligned} \|g_\epsilon - f\|_p &= \left\{ \int_{\mathbb{R}} |g_\epsilon(x) - f(x)|^p \, dx \right\}^{1/p} = \left\{ \int_{[-\beta,\beta]} |g_\epsilon(x) - f(x)|^p \, dx \right\}^{1/p} \\ &\leq \sup_{x\in[-\beta,\beta]} |g_\epsilon(x) - f(x)|(2\beta)^{1/p} < \epsilon. \end{aligned}$$

Now, the set of rational numbers is countable, and this can be used to establish that the set $P_{\mathbb{Q}}[-\beta, \beta]$ is also countable. If we define the set $\Gamma = \{g : g = q \cdot \chi_{[-\beta,\beta]}$ for some $q \in P_{\mathbb{Q}}[-\beta, \beta]$ and some $\beta = 1, 2, \ldots\}$, then it can be shown that Γ is a countable set in $C_0^\infty(\mathbb{R})$. The above inequality indicates that Γ is dense in $C_0^\infty(\mathbb{R})$, and hence it is dense in $L^p(\mathbb{R})$. Thus, $L^p(\mathbb{R})$ contains a countable dense set.

8.7 Complex L^p Spaces

In this section we pause to make a modest generalization of L^p spaces to include complex-valued functions of a real variable. A fuller treatment of these spaces can be found in [17]. Let $I \subseteq \mathbb{R}$ be some interval and let $f : I \to \mathbb{C}$ be a complex-valued function defined on I. The function f can always be decomposed into the form $f = \text{Re} f + i\text{Im} f$, where $\text{Re} f$ and $\text{Im} f$ are real-valued functions. If $\text{Re} f$ and $\text{Im} f$ are integrable functions, then the Lebesgue integral of f is defined as

$$\int_I f(t) \, dt = \int_I \text{Re} f(t) \, dt + i \int_I \text{Im} f(t) \, dt.$$

We wish to construct spaces analogous to $\Lambda^p(I)$ and $L^p(I)$. The natural generalization of the norm function is the function $\|\cdot\|_p$ defined by

$$\|f\|_p = \left\{ \int_I |f(t)|^p \, dt \right\}^{1/p},$$

where for $\bar{f} = \text{Re} f - i\text{Im} f$ (the complex conjugate), $|f|^2 = f\bar{f}$. The next theorem shows that $|f|^p$ is integrable if and only if $|\text{Re} f|^p$ and $|\text{Im} f|^p$ are both integrable. The proof rests on the inequalities $|\text{Re} f| \leq |f|, |\text{Im} f| \leq |f|$, and $|f| \leq |\text{Re} f| + |\text{Im} f|$, and is left as an exercise.

Theorem 8.7.1

Let $f : I \to \mathbb{C}$ be a complex-valued function defined on an interval $I \subseteq \mathbb{R}$ and suppose $f = \text{Re} f + i\text{Im} f$, where $\text{Re} f$ and $\text{Im} f$ are real-valued measurable functions. Then, for $1 \leq p < \infty$,

$$\int_I |f(t)|^p \, dt < \infty$$

if and only if

$$\int_I |\text{Re} f(t)|^p \, dt < \infty \quad \text{and} \quad \int_I |\text{Im} f(t)|^p \, dt < \infty.$$

Let $\Lambda^p_{\mathbb{C}}(I)$ denote the set of functions $f : I \to \mathbb{C}$ such that $\|f\|_p < \infty$. The above result asserts that $f \in \Lambda^p_{\mathbb{C}}(I)$ if and only if $\text{Re} f \in \Lambda^p(I)$ and $\text{Im} f \in \Lambda^p(I)$.

For $1 < p < \infty$ it can be shown that the function $\| \cdot \|_p$ satisfies the Hölder and Minkowski inequalities. In particular, the sets $\Lambda^p_{\mathbb{C}}(I)$ are complex vector spaces and $\| \cdot \|_p$ is a seminorm. As with the $\Lambda^p(I)$ sets, we can form a normed vector space by partitioning $\Lambda^p_{\mathbb{C}}(I)$ into equivalence classes. Two functions belong to the same equivalence class if $|f - g| = 0$ a.e. Let $L^p_{\mathbb{C}}(I)$ denote the resulting set of equivalence classes. The spaces $L^p_{\mathbb{C}}(I)$ share the same properties as the $L^p(I)$ spaces. For example, if $1 \leq p < \infty$, the complex vector space $L^p_{\mathbb{C}}(I)$ is a Banach space. If p and q are conjugate exponents and $p > 1$, then the dual space of $L^p_{\mathbb{C}}(I)$ is the space $L^q_{\mathbb{C}}(I)$. If J is a bounded linear functional from $L^p_{\mathbb{C}}(I)$ to $\mathbb{C}$, then there is a unique $g \in L^q_{\mathbb{C}}(I)$ such that

$$J(f) = \int_I f(t)\bar{g}(t) \, dt$$

for all $f \in L^p_{\mathbb{C}}(I)$. The space $L^\infty_{\mathbb{C}}(I)$ can be defined in a manner analogous to that used to define $L^\infty(I)$, and this space is the dual of $L^1_{\mathbb{C}}(I)$.

The $L^p_{\mathbb{C}}(I)$ spaces often arise in applications involving line integrals in the complex plane. Recall that if γ is some curve in $\mathbb{C}$ represented parametrically by some piecewise smooth function $z(t) = x(t) + iy(t)$ for $t \in [t_0, t_1]$, then the line integral of a function F defined on γ is given by

$$\int_\gamma F(z)\,dz = \int_{t_0}^{t_1} F(z(t))z'(t)\,dt.$$

The definition of a line integral can be extended using the Lebesgue integral in an obvious way, and thus line integrals can be defined for a more general class of functions. Note that a curve γ may be parametrized any number of ways. For example, the variable t may be replaced by any smooth function $g : [s_0, s_1] \to [t_0, t_1]$ to give a new parametrization $\hat{z}(s) = z(g(s))$, provided that $g'(s) \neq 0$ for all $s \in [s_0, s_1]$. If $g'(s) > 0$ in the interval $[s_0, s_1]$ then the parametrization is orientation preserving, and it is well known that the value of the line integral in the Riemann setting is invariant under smooth orientation preserving reparametrizations. In the Lebesgue setting, Theorem 6.2.1 ensures that this is also the case. Thus, provided that the orientation of γ is specified, we can use the notation $\int_\gamma F(z)dz$ without ambiguity. In this context we may also use the notation $L^p_{\mathbb{C}}(\gamma)$ unless a specific parametrization of γ is required.

A particularly interesting case is that in which γ is a simple closed curve (no self-intersections) and $F(z)$ is holomorphic (analytic) in the region enclosed by γ but not necessarily on γ itself. In the next section, we investigate a class of spaces known as Hardy spaces and show that the $L^p_{\mathbb{C}}(I)$ spaces are closely related to spaces of holomorphic functions.

8.8 The Hardy Spaces H^p

Hardy spaces are complex normed vector spaces of functions holomorphic on a region of the complex plane. These spaces are closely related to the Lebesgue spaces and share the same properties. The functions in a Hardy space are quite different in nature from elements in an $L^p_{\mathbb{C}}(I)$ space. Aside from the fact that elements in the latter set are equivalence classes, the underlying functions in an $L^p_{\mathbb{C}}(I)$ space need not be continuous. In contrast, the functions in a Hardy space are infinitely differentiable. That these spaces should be so closely linked is remarkable. Hardy spaces play an important role in complex function theory and harmonic analysis. These spaces also arise in applications such as control theory. The discussion in this section presumes some knowledge of complex analysis, but it is primarily a descriptive account and no proofs are given. For a basic reference on complex analysis the reader is directed to [2], [10], or [39]. A much fuller discussion of Hardy spaces along with the proofs of various results in this section can be found in [20], [21], and [37].

Let $\Omega \subseteq \mathbb{C}$ denote a region (i.e., a nonempty, connected open set) and let $\mathcal{A}(\Omega)$ denote the set of functions holomorphic on Ω. It is clear from the elementary properties of holomorphic functions that $\mathcal{A}(\Omega)$ is a vector space. A candidate for a norm on $\mathcal{A}(\Omega)$ is the function $\|\cdot\|_\infty$ defined by

$$\|f\|_\infty = \sup_{z\in\Omega} |f(z)|.$$

Although this function satisfies the conditions for a norm such as the triangle inequality, it is not finite for all $f \in \mathcal{A}(\Omega)$. There are functions in $\mathcal{A}(\Omega)$ with singularities on the boundary $\partial\Omega$ of Ω. If the set $\mathcal{A}(\Omega)$ is restricted to the functions f for which $\|f\|_\infty < \infty$, then the resulting space is a normed vector space. Let

$$H^\infty(\Omega) = \{f \in \mathcal{A}(\Omega) : \|f\|_\infty < \infty\}.$$

The set $H^\infty(\Omega)$ is still a vector space, and now $\|\cdot\|_\infty$ is a norm on it. In fact, the space $(H^\infty(\Omega), \|\cdot\|_\infty)$ is a Banach space.

The value of $\|f\|_\infty$ for a given $f \in H^\infty(\Omega)$ is less tractable than the analogous function defined for $C[a,b]$ (Example 8-1-5) owing to the two-dimensional nature of the region Ω and the fact that Ω is an *open* set. A fundamental result in complex analysis known as the *maximum modulus principle*, however, mitigates these problems. Let $\Omega \subseteq \mathbb{C}$ be a *bounded* region with boundary $\partial\Omega$ and suppose that $f \in \mathcal{A}(\Omega)$ and also that f is continuous in the set $\bar{\Omega} = \Omega \cup \partial\Omega$. Under these conditions, the maximum modulus theorem implies that the function $|f|$ assumes its maximum value on the boundary $\partial\Omega$ and that a nonconstant function in $\mathcal{A}(\Omega)$ cannot have local maxima for $|f|$ in Ω.

The maximum modulus principle cannot be applied directly to all the functions in $H^\infty(\Omega)$, since functions in this set need not be continuous on $\Omega \cup \partial\Omega$. Nonetheless, the norm $\|f\|_\infty$ of a function in $H^\infty(\Omega)$ reflects the values of $|f(z)|$ as z approaches the boundary of the region, and this suggests that the norm can be calculated using a limiting argument.

The *Riemann mapping theorem* implies that arbitrary regions such as S can be conformally mapped to a unit disk in the complex plane. We can thus limit our investigations mostly to the study of functions holomorphic on the unit disk centred at the origin. Let $D(0;r) = \{z \in \mathbb{C} : |z| < r\}$, $\bar{D}(0;r) = \{z \in \mathbb{C} : |z| \leq r\}$, $\partial D_r = \{z \in \mathbb{C} : |z| = r\}$, and for simplicity, denote $D(0;1)$, $\bar{D}(0;1)$, ∂D_1 by D, $\bar{D}$, and ∂D, respectively. If $0 < r < 1$, then, taking $\Omega = D(0;r)$, the maximum modulus principle implies that

$$\sup_{z \in D(0;r)} |f(z)| = \sup_{z \in \partial D_r} |f(z)|.$$

The function M_∞ is defined by

$$M_\infty(r,f) = \sup_{\phi \in [0,2\pi]} |f(re^{i\phi})|,$$

and it is evident from the maximum modulus principle that for a fixed $f \in H^\infty(D)$ the function M_∞ is monotonic increasing in r and that

$$\sup_{z \in D(0;r)} |f(z)| = M_\infty(r,f).$$

We thus expect that for any $f \in H^\infty(D)$,

$$\lim_{r \to 1^-} M_\infty(r,f) = \|f\|_\infty,$$

and in fact, this relationship can be proved formally.

The function M_∞ suggests a generalization for norms analogous to the L^p norms. Let M_p be the function defined by

$$M_p(r,f) = \left\{ \frac{1}{2\pi} \int_{[0,2\pi]} |f(re^{i\phi})|^p \, d\phi \right\}^{1/p},$$

and let $H^p(D)$ be defined as

$$H^p(D) = \{f \in \mathcal{A}(D) : \sup_{r<1} M_p(r, f) < \infty\}.$$

It can be shown that for $1 \leq p < \infty$ the M_p functions satisfy Hölder- and Minkowski-type inequalities. Specifically, if p and q are conjugate exponents with $p > 1$, then for any functions $f \in H^p(D)$, $g \in H^q(D)$,

$$M_1(r, fg) \leq M_p(r, f) M_q(r, g),$$

and if $h \in H^p(D)$, then

$$M_p(r, f + h) \leq M_p(r, f) + M_p(r, h).$$

If $p \geq 1$, the function $\| \cdot \|_p$ defined by

$$\|f\|_p = \sup_{r<1} M_p(r, f)$$

is therefore a norm on $H^p(D)$. (Note that we need not form equivalence classes because the functions involved are holomorphic, and thus if $f = g$ a.e., then $f = g$.) As with $M_\infty(r, f)$, it can be shown that for all $f \in H^p(D)$,

$$\lim_{r \to 1^-} M_p(r, f) = \|f\|_p.$$

The normed vector spaces $(H^p(D), \| \cdot \|_p)$ are called **Hardy** or H^p **spaces**. As with the Lebesgue spaces, we refer to the normed vector space $(H^p(D), \| \cdot \|_p)$ as $H^p(D)$. The Hardy spaces share the same properties as the Lebesgue spaces. The next theorem gives a sample of some of these properties.

Theorem 8.8.1

(i) **Hölder's Inequality:** *If $1 < p < \infty$ and q is the conjugate exponent of p, then for any $f \in H^p(D)$, $g \in H^q(D)$ we have that $fg \in H^1(D)$ and*

$$\|fg\|_1 \leq \|f\|_p \|g\|_q.$$

(ii) **Minkowski's Inequality:** *For any $f, g \in H^p(D)$ with $p \geq 1$,*

$$\|f + g\|_p \leq \|f\|_p + \|g\|_p.$$

(iii) **Completeness:** *The normed vector spaces $(H^p(D), \| \cdot \|_p)$ are Banach spaces.*

(iv) **Inclusion:** *For any* $1 \leq p_1 \leq p_2 < \infty$,

$$H^{\infty}(D) \subseteq H^{p_2}(D) \subseteq H^{p_2}(D).$$

The two inequalities in the above theorem are direct consequences of the corresponding inequalities involving the M_p functions. A proof of completeness can be found in [21]. The inclusion result is established using the same approach as used to prove a similar result in Theorem 8.5.4.

It is not a coincidence that the H^p spaces mimic the L^p spaces. These spaces are closely related through the "boundary values" of the functions in $H^p(D)$. Cauchy's integral formula is a remarkable manifestation that a holomorphic function is determined by its boundary values. Recall that for any $f \in \mathcal{A}(\bar{D}(0;r))$, $r > 0$, the Cauchy integral formula states that

$$f(z) = \frac{1}{2\pi i} \int_{\partial D_r} \frac{f(w)}{w - z}\, dw$$

for any $z \in D(0;r)$. Here, the circle ∂D_r is oriented anticlockwise. The value of f on the boundary ∂D_r thus determines the function f uniquely in the interior $D(0;r)$. We cannot apply Cauchy's integral formula directly on the boundary of D because f need not be holomorphic on ∂D, but we can still apply the formula on any boundary ∂D_r when $0 < r < 1$. This suggests that any function $f \in H^p(D)$ can be "identified" with a function $\tilde{f}$ defined on ∂D a.e. by some limiting process as $r \to 1^-$.

Let

$$f_r(z) = f(rz).$$

If $0 < r < 1$, then $f_r(z)$ is holomorphic on $\bar{D}$ and the Cauchy integral formula implies that

$$f_r(z) = \frac{1}{2\pi i} \int_{\partial D} \frac{f_r(\xi)}{\xi - z}\, d\xi.$$

Intuitively, we should be able to define a function $\tilde{f} : \partial D \to \mathbb{C}$ by considering the function $\lim_{r \to 1^-} f_r$. The next result shows that this approach does lead to an isometry from the $H^p(D)$ spaces to the $L^p_{\mathbb{C}}(\partial D)$ spaces. The notation for the norms for these spaces is the same by convention. Here, we shall use the notation $\|\cdot\|_{H^p}$, and $\|\cdot\|_{L^p}$ to distinguish the norm for $H^p(D)$ from the norm for $L^p_{\mathbb{C}}(\partial D)$.

Theorem 8.8.2

Let $1 \leq p < \infty$ and suppose that $f \in H^p(D)$. Then there is an element $[\tilde{f}] \in L^p_{\mathbb{C}}(\partial D)$ such that:

(i) $\|\tilde{f}\|_{L^p} = \|f\|_{H^p}$.

(ii) $\lim_{r\to 1^-} \|\tilde{f} - f_r\|_{L^p} = 0$.

(iii)

$$f(z) = \frac{1}{2\pi i}\int_{\partial D} \frac{\tilde{f}(\xi)}{\xi - z}\, d\xi.$$

The above result indicates that there is an isometry from $H^p(D)$ to $L^p_{\mathbb{C}}(\partial D)$. These spaces are not isometric, since the set $L^p_{\mathbb{C}}(\partial D)$ contains elements with no corresponding members in $H^p(D)$. Let $\mathcal{H}^p(\partial D) \subseteq L^p_{\mathbb{C}}(\partial D)$ denote the set of elements $[\tilde{f}] \in L^p_{\mathbb{C}}(\partial D)$ such that $\tilde{f}$ corresponds to some function $f \in H^p(D)$. The complex normed vector spaces $(\mathcal{H}^p(\partial D), \|\cdot\|_{L^p})$ and $(H^p(D), \|\cdot\|_{H^p})$ are isometric. Now, it can be shown that the space $(\mathcal{H}^p(\partial D), \|\cdot\|_{L^p})$ is a *closed* subspace of the Banach space $(L^p_{\mathbb{C}}(\partial D), \|\cdot\|_{L^p})$, and this means that $(\mathcal{H}^p(\partial D), \|\cdot\|_{L^p})$ is also a Banach space. Since $(\mathcal{H}^p(\partial D), \|\cdot\|_{L^p})$ is isometric to $(H^p(D), \|\cdot\|_{H^p})$, the latter space must also be a Banach space. Properties of the $H^p(D)$ spaces such as completeness are generally proved by establishing the analogous results for the $\mathcal{H}^p(\partial D)$ spaces, and in this sense the Hardy spaces "inherit" the qualities of the L^p spaces.

The $H^p(D)$ spaces can be generalized to sets of functions holomorphic in arbitrary regions or the complex plane via conformal transformations. Perhaps the most frequently encountered Hardy spaces aside from the $H^p(D)$ spaces are the spaces of functions holomorphic on a half-plane. Specifically, the Hardy spaces of functions analytic on the right half-plane $\Pi_0 = \{z \in \mathbb{C} : \operatorname{Re} z > 0\}$ are of interest owing to (among other things) their connection with the Laplace transform. The Hardy spaces in the half-plane share most of the properties of the $H^p(D)$ spaces. Indeed these properties are sometimes established by using a Möbius transformation such as

$$z = \frac{w-1}{w+1}.$$

This transformation maps points $z \in D$ to points $w \in \Pi_0$; the circle ∂D is mapped to the imaginary axis. If $g \in \mathcal{A}(D)$, then the function f defined by

$$f(w) = g\left(\frac{w-1}{w+1}\right) \tag{8.2}$$

is in $\mathcal{A}(\Pi_0)$, and in this manner we can "transport" some of the properties of the $H^p(D)$ spaces. We return to this comment at the end of the section.

Let

$$M_p(x,f) = \left\{\int_{\mathbb{R}} |f(x+iy)|^p \, dy\right\}^{1/p},$$

and define the function $\|\cdot\|_p$ by

$$\|f\|_p = \sup_{x>0} M_p(x,f).$$

The Hardy spaces $H^p(\Pi_0)$ are defined as

$$H^p(\Pi_0) = \{f \in \mathcal{A}(\Pi_0) : \|f\|_p < \infty\}.$$

In contrast with the set D, the set Π_0 is not bounded in the complex plane, and this complicates matters because along any line $I_x = \{z \in \mathbb{C} : \operatorname{Re} z = x, x > 0\}$ the function f_x defined by

$$f_x(y) = f(x+iy)$$

must be in $L^p(I_x)$. Moreover, we cannot appeal directly to the maximum modulus principle to evaluate $\|f\|_p$ because the region is unbounded and the function may have a singularity at the point at infinity. The bound for $M_p(x,f)$ for large x is as important as that for small x for the general $f \in \mathcal{A}(\Pi_0)$. It turns out, however, that the requirement that $\|f\|_p < \infty$ forces f to tend uniformly to zero as z tends toward the point at infinity along any path inside any *fixed* half-plane of the form $\Pi_\delta = \{z \in \mathbb{C} : \operatorname{Re} z \geq \delta > 0\}$. It can thus be shown that $M_p(x,f)$ is a decreasing function of x and that

$$\lim_{x\to 0^+} M_p(x,f) = \|f\|_p.$$

As with the $H^p(D)$ spaces, the boundary values of functions in $H^p(\Pi_0)$ can be mapped to elements in an L^p space. The

Lebesgue space associated with $H^p(\Pi_0)$ is the space $L^p_{\mathbb{C}}(\mathbb{R})$, and the corresponding Cauchy (Poisson) integral for this case is

$$f_x(y) = \frac{x}{\pi}\int_{\mathbb{R}} \frac{\tilde{f}(it)}{x^2 + (y-t)^2}\, dt.$$

For each $x > 0$ the function f_x is in $L^p_{\mathbb{C}}(\mathbb{R})$, and

$$\|f_x - \tilde{f}\|_{L^p} \to 0$$

as $x \to 0^+$. Moreover, we have that $\|f\|_{H^p} = \|\tilde{f}\|_{L^p}$.

As remarked earlier, the sets D and Π_0 are related by a Möbius transformation, and we can expect that the sets of functions in one space can be used to generate the functions in the other space. If $g \in H^p(D)$, then the function f as defined in equation (8.2) is certainly in $\mathcal{A}(\Pi_0)$, but it is not clear whether $f \in H^p(\Pi_0)$ and whether all the functions in $H^p(\Pi_0)$ can be generated by functions in $H^p(D)$. The final result of this section gives a concrete connection between the two spaces (cf. [21]).

Theorem 8.8.3
Let $g \in \mathcal{A}(D)$ and define f as in equation (8.2). If $p \geq 1$, then $f \in H^p(\Pi_0)$ if and only if there is a function $G \in H^p(D)$ such that

$$g(z) = (1-z)^{2/p}G(z).$$

Equivalently, the function g is in $H^p(D)$ if and only if there is a function $F \in H^p(\Pi_0)$ such that

$$f(w) = (1+w)^{2/p}F(w).$$

8.9 Sobolev Spaces $W^{k,p}$

Sobolev spaces are another class of function spaces based on the Lebesgue integral. These function spaces have found widespread applications in differential equations and feature norms that not only "measure" the modulus of a function but also the modulus of its derivatives in an L^p setting. In this section we give a brief glimpse of these important spaces and hopefully whet the reader's appetite for a more serious study. A detailed account of the theory can be found in [1].

Let $C^1(I)$ denote the set of functions f mapping the interval $I \subseteq \mathbb{R}$ to $\mathbb{R}$ such that f' exists and is continuous for all $x \in I$. The set $C^1(I)$ forms a vector space, and the function $\|\cdot\|_{1,\infty}$ defined by

$$\|f\|_{1,\infty} = \sup_{x\in I} |f(x)| + \sup_{x\in I} |f'(x)|$$

is a norm. The space $(C^1(I), \|\cdot\|_{1,\infty})$ can be shown to be complete. More generally, we can define functions $\|\cdot\|_{1,p}$ by

$$\|f\|_{1,p} = \left\{\int_I |f(x)|^p\,dx\right\}^{1/p} + \left\{\int_I |f'(x)|^p\,dx\right\}^{1/p} = \|f\|_p + \|f'\|_p,$$

and investigate the corresponding normed vector spaces $(C^1(I), \|f\|_{1,p})$. For $1 \leq p < \infty$ we know that $(C(I), \|\cdot\|_p)$ is not complete. The completion of this space essentially led us to use the Lebesgue integral instead of the Riemann integral and "expand" the vector space to $L^p(I)$. Now we know that the normed vector space $(C^1(I), \|f\|_{1,p})$ has a completion, but it is not clear what new concepts will be involved in finding it. Evidently, we need to enlarge our vector space to include functions that have a first derivative in $L^p(I)$, but is this adequate? That f must have a derivative in the classical sense requires f to be continuous, and this seems a strong restriction in the Lebesgue setting.

The completion of spaces such as $(C^1(I), \|\cdot\|_{1,p})$ leads to the concept of distributions (generalized functions) and generalized derivatives. A discussion about these functions would lead us far astray. Suffice it to say that spaces like $(C^1(I), \|\cdot\|_{1,p})$ can be completed, but the "cost" is the introduction of an entire new class of objects that are not functions in the classical sense. Once again, the completion of a space yields a new mathematical concept.

The **Sobolev spaces** are the completions of spaces like $(C^1(I), \|f\|_{1,p})$. The completion of $(C^1(I), \|\cdot\|_{1,p})$ is denoted by $(W^{1,p}(I), \|\cdot\|_{1,p})$ or simply $W^{1,p}(I)$. If $C^k(I)$ denotes the set of functions with kth-order continuous derivatives, and $\|\cdot\|_{k,p}$ is defined by

$$\|f\|_{k,p} = \sum_{j=0}^{k} \|f^{(j)}\|_p,$$

then the Sobolev space $(W^{k,p}(I), \|\cdot\|_{k,p})$ is defined as the completion of the space $(C^k(I), \|\cdot\|_{k,p})$.

9 CHAPTER Hilbert Spaces and L^2

9.1 Hilbert Spaces

Hilbert spaces are a special class of Banach spaces. Hilbert spaces are simpler than Banach spaces owing to an additional structure called an inner product. These spaces play a significant role in functional analysis and have found widespread use in applied mathematics. We shall see at the end of this section that the Lebesgue space L^2 (and its complex relative H^2) is a Hilbert space. In this and the next section, we introduce some basic definitions and facts concerning Hilbert spaces of immediate interest to our discussion of the space L^2. Further details and proofs of the results presented in these sections can be found in most books on functional analysis, e.g., [25].

Let X be a real or complex vector space. An **inner product** is a scalar-valued function $\langle \cdot, \cdot \rangle$ on $X \times X$ such that for any $f, g, h \in X$ and any scalar α the following conditions hold:

(i) $\langle f, f \rangle \geq 0$;

(ii) $\langle f, f \rangle = 0$ if and only if $f = \mathbf{0}$;

(iii) $\langle f + g, h \rangle = \langle f, h \rangle + \langle g, h \rangle$;

(iv) $\langle f, g \rangle = \overline{\langle g, f \rangle}$;

(v) $\langle \alpha f, g \rangle = \alpha \langle f, g \rangle$.

A vector space X equipped with an inner product $\langle\cdot,\cdot\rangle$ is called an **inner product space** and denoted by $(X, \langle\cdot,\cdot\rangle)$. Note that in general, $\langle f, g\rangle$ is a complex number; however, condition (i) indicates that $\langle f, f\rangle$ is always a real nonnegative number. Note also that conditions (iii) and (iv) imply that

$$\langle f, g+h\rangle = \langle f, g\rangle + \langle f, h\rangle.$$

In general, $\langle f, \beta g\rangle = \overline{\beta}\langle f, g\rangle \neq \beta\langle f, g\rangle$, and consequently the inner product is not in general a bilinear function. The special case arises frequently in applications that X is a real vector space and $\langle\cdot,\cdot\rangle$ is real-valued. These spaces are referred to as **real inner product spaces**, and for these spaces the inner product is a bilinear function.

Example 9-1-1: $\mathbb{C}^n$ and $\mathbb{R}^n$

Let $X = \mathbb{C}^n$ and for any $\mathbf{z} = (z_1, z_2, \ldots, z_n), \mathbf{w} = (w_1, w_2, \ldots, w_n) \in \mathbb{C}^n$ define $\langle\cdot,\cdot\rangle$ by

$$\langle \mathbf{z}, \mathbf{w}\rangle = \sum_{j=1}^{n} z_j \overline{w}_j.$$

Then it is straightforward to verify that $\langle\cdot,\cdot\rangle$ is an inner product on $\mathbb{C}^n$. Similarly, if $X = \mathbb{R}^n$ and for any $\mathbf{x} = (x_1, x_2, \ldots, x_n), \mathbf{y} = (y_1, y_2, \ldots, y_n) \in \mathbb{R}^n$ the function $\langle\cdot,\cdot\rangle$ is defined by

$$\langle \mathbf{x}, \mathbf{y}\rangle = \sum_{j=1}^{n} x_j y_j,$$

then $\langle\cdot,\cdot\rangle$ defines an inner product on $\mathbb{R}^n$. The definition of the inner product is modeled after the familiar inner product (dot product) defined for $\mathbb{R}^n$.

Example 9-1-2: ℓ^2

Let ℓ^2 denote the set of complex sequences $\{a_n\}$ such that the series $\sum_{n=1}^{\infty} |a_n|^2$ is convergent. If addition and scalar multiplication are defined the same way as for the space ℓ^1 in Example 9-1-3, then ℓ^2 is a vector space. Suppose that $\mathbf{a} = \{a_n\}, \mathbf{b} = \{b_n\} \in \ell^2$, and let $c_n = \max(a_n, b_n)$. Then the series $\sum_{n=1}^{\infty} |c_n|^2$ is convergent, and hence the series $\sum_{n=1}^{\infty} a_n \overline{b}_n$ is absolutely convergent. An inner product on this vector space is defined by

$$\langle \mathbf{a}, \mathbf{b} \rangle = \sum_{n=1}^{\infty} a_n \overline{b}_n.$$

Let $(X, \langle \cdot, \cdot \rangle)$ be an inner product space and let $\| \cdot \| : X \to \mathbb{R}$ be the function defined by

$$\|f\| = \sqrt{\langle f, f \rangle}.$$

We will show that $\| \cdot \|$ as defined above is a norm hence justifying our notation. The conditions for the inner product ensure that $\| \cdot \|$ meets the requirements for a norm except perhaps the triangle inequality. In order to establish the triangle inequality we need the following result, which is of interest in its own right:

Theorem 9.1.1 (Schwarz's Inequality)
Let $(X, \langle \cdot, \cdot \rangle)$ be an inner product space. If $f, g \in X$, then

$$|\langle f, g \rangle| \leq \|f\| \|g\|.$$

The proof of this inequality is left as an exercise. Now,

$$\begin{aligned} \|f+g\|^2 &= \langle f+g, f+g \rangle \\ &= \langle f, f \rangle + \overline{\langle f, g \rangle} + \langle f, g \rangle + \langle g, g \rangle \\ &= \|f\|^2 + 2\mathrm{Re}\, \langle f, g \rangle + \|g\|^2 \\ &\leq \|f\|^2 + 2|\langle f, g \rangle| + \|g\|^2, \end{aligned}$$

and Schwarz's inequality implies that

$$\begin{aligned} \|f+g\|^2 &\leq \|f\|^2 + 2\|f\| \|g\| + \|g\|^2 \\ &= (\|f\| + \|g\|)^2. \end{aligned}$$

Thus,

$$\|f+g\| \leq \|f\| + \|g\|,$$

and hence $\| \cdot \|$ defines a norm on X.

Given any inner product space $(X, \langle \cdot, \cdot \rangle)$ we can construct a normed vector space $(X, \| \cdot \|)$. The function $\| \cdot \|$ is called the **norm induced by the inner product**. The normed vector space may or may not be complete. If $(X, \| \cdot \|)$ is a Banach space, then the inner product space $(X, \langle \cdot, \cdot \rangle)$ is called a **Hilbert space**. A Hilbert space is thus an inner product space that is complete in the norm induced by the inner product. The inner product spaces $(\mathbb{R}^n, \langle \cdot, \cdot \rangle)$ and $(\mathbb{C}^n, \langle \cdot, \cdot \rangle)$ are examples of finite-dimensional Hilbert spaces. Although we do not show it here, the inner product space $(\ell^2, \langle \cdot, \cdot \rangle)$ of Example 9-1-2 is an infinite-dimensional Hilbert space.

It is of interest to enquire whether a given normed space $(X, \| \cdot \|)$ can be identified with an inner product space $(X, \langle \cdot, \cdot \rangle)$. In other words, given a norm $\| \cdot \|$ on X can an inner product $\langle \cdot, \cdot \rangle$ be defined on X such that $\| \cdot \|$ corresponds to the norm induced by $\langle \cdot, \cdot \rangle$? Suppose that the norm $\| \cdot \|$ can be obtained from some inner product $\langle \cdot, \cdot \rangle$. Then for any $f, g \in X$,

$$\begin{aligned} \|f+g\|^2 + \|f-g\|^2 &= \langle f+g, f+g \rangle + \langle f-g, f-g \rangle \\ &= \langle f, f \rangle + \langle f, g \rangle + \langle g, f \rangle + \langle g, g \rangle \\ &\quad + \langle f, f \rangle + \langle f, -g \rangle + \langle -g, f \rangle + \langle g, g \rangle \\ &= 2\left(\|f\|^2 + \|g\|^2\right); \end{aligned}$$

thus, if a norm $\| \cdot \|$ can be obtained by some inner product, then it must satisfy the equation

$$\|f+g\|^2 + \|f-g\|^2 = 2\left(\|f\|^2 + \|g\|^2\right).$$

This condition is also sufficient. This equation is called the **parallelogram equality**. The name comes from an elementary relation in plane geometry. If $\mathbf{x}$ and $\mathbf{y}$ are two vectors in $\mathbb{R}^2$ that are not parallel, then they can be used to define a parallelogram. Here, the quantities $\|\mathbf{x}\|$ and $\|\mathbf{y}\|$ correspond to the lengths of the vectors $\mathbf{x}$ and $\mathbf{y}$ respectively and hence the lengths of the sides of the parallelogram. The quantities $\|\mathbf{x}+\mathbf{y}\|$ and $\|x-y\|$ correspond to the lengths of the diagonals. The parallelogram equality is useful for (among other things) showing that certain norms cannot be obtained from an inner product.

Example 9-1-3:
Consider the normed vector space $(C[a,b], \| \cdot \|_\infty)$ defined in Example 8-1-5. We shall use the parallelogram equality to show that the norm $\| \cdot \|_\infty$ on $C[a,b]$ cannot be obtained by an inner product. Suppose, for contradiction, that the norm $\| \cdot \|_\infty$ can be obtained by an inner product. Then the parallelogram equality must be satisfied for any

choice of $f, g \in C[a, b]$. Let f and g be the functions defined by

$$f(x) = 1, \quad g(x) = \frac{x-a}{b-a}.$$

Now, $\|f\|_\infty = 1$, $\|g\|_\infty = 1$,

$$\|f+g\|_\infty = \sup_{x\in[a,b]} \left|1 + \frac{x-a}{b-a}\right| = 2,$$

and

$$\|f-g\|_\infty = \sup_{x\in[a,b]} \left|1 - \frac{x-a}{b-a}\right| = 1;$$

consequently,

$$\|f+g\|_\infty^2 + \|f-g\|_\infty^2 = 4+1 = 5,$$

and

$$2\left(\|f\|_\infty^2 + \|g\|_\infty^2\right) = 4.$$

As the parallelogram equality is not satisfied for these functions, the normed vector space $(C[a, b], \|\cdot\|_\infty)$ cannot be obtained from an inner product space.

The parallelogram equality can be used to show that the only L^p space that might arise from an inner product space is L^2.

Theorem 9.1.2
The only Lebesgue norm $\|\cdot\|_p$ that can be obtained from an inner product is the L^2 norm $\|\cdot\|_2$.

Proof We first establish the result for $L^p[-1, 1]$ spaces. A modest change of the proof leads to the result for general $L^p[a, b]$ spaces and $L^p(I)$ spaces where I is an unbounded interval.

Suppose that the norm $\|\cdot\|_p$ on $L^p[-1, 1]$ can be obtained from an inner product. Then the parallelogram equality indicates that

$$\|f+g\|_p^2 + \|f-g\|_p^2 = 2\left(\|f\|_p^2 + \|g\|_p^2\right)$$

for all $f, g \in L^p[-1, 1]$. Consider the functions f and g defined by

$$f(x) = 1+x, \quad g(x) = 1-x.$$

These functions are (in equivalence classes) in $L^p[-1,1]$ for all $p \geq 1$. Now,

$$\|f\|_p^p = \int_{[-1,1]} |1+x|^p \, dx = \int_{-1}^{1} (1+x)^p \, dx = \frac{2^{p+1}}{p+1},$$

$$\|g\|_p^p = \int_{[-1,1]} |1-x|^p \, dx = \int_{-1}^{1} (1-x)^p \, dx = \frac{2^{p+1}}{p+1},$$

and

$$\|f+g\|_p^p = \int_{[-1,1]} |1+x+(1-x)|^p \, dx = \int_{-1}^{1} 2^p \, dx = 2^{p+1},$$

$$\|f-g\|_p^p = \int_{[-1,1]} |1+x-(1-x)|^p \, dx = \int_{-1}^{1} |2x|^p \, dx = 2^{p+1} \int_0^1 x^p \, dx = \frac{2^{p+1}}{p+1}.$$

The parallelogram equality implies that

$$\left(2^{p+1}\right)^{2/p} + \left(\frac{2^{p+1}}{p+1}\right)^{2/p} = 2\left\{\left(\frac{2^{p+1}}{p+1}\right)^{2/p} + \left(\frac{2^{p+1}}{p+1}\right)^{2/p}\right\},$$

i.e.,

$$(p+1)^{2/p} - 3 = 0.$$

Note that $p = 2$ is a solution to this equation. We will show that this is the only solution for $p \geq 1$. Let

$$Q(p) = (p+1)^{2/p} - 3.$$

Then

$$Q'(p) = \frac{2}{p^2(p+1)}(p+1)^{2/p} S(p),$$

where

$$S(p) = p - (p+1)\log(p+1).$$

Now,

$$S'(p) = 1 - \frac{p+1}{p+1} - \log(p+1) = -\log(p+1) < 0,$$

and thus S is a monotonic strictly decreasing function of p. Since $S(1) = 1 - 2\log 2 < 0$, it follows that $S(p) < 0$ for all $p \geq 1$ and consequently that $Q'(p) < 0$ for all $p \geq 1$. This means that Q is a monotonic strictly decreasing function of p, and therefore the equation $Q(p) = 0$ can have at most one solution. Therefore, the parallelogram equality is satisfied only if $p = 2$.

A slight modification of the functions f, g for the general interval $[a, b]$ leads to the proof of the result for the $L^p[a, b]$ spaces. If I is not a bounded interval, then a suitable restriction of the functions can be used to establish the result for the $L^p(I)$ spaces. For example, suppose that $I = (-\infty, \infty)$. Then we can choose the functions

$$f(x) = \begin{cases} 1+x, & \text{if } x \in [-1, 1], \\ 0, & \text{otherwise,} \end{cases}$$

$$g(x) = \begin{cases} 1-x, & \text{if } x \in [-1, 1], \\ 0, & \text{otherwise,} \end{cases}$$

and the proof then is exactly the same. □

It is not difficult to see that for any interval I, $L^2(I)$ is a Hilbert space. We know from Chapter 8 that this space is complete with respect to the $\|\cdot\|_2$ norm, and the inner product $\langle\cdot,\cdot\rangle$ defined by

$$\langle f, g\rangle = \int_I f(x)g(x)\,dx$$

for all $f, g \in L^2(I)$ induces this norm. The complex space $L^2_{\mathbb{C}}(I)$ is also a Hilbert space with an inner product $\langle\cdot,\cdot\rangle$ defined by

$$\langle f, g\rangle = \int_I f(t)\overline{g(t)}\,dt.$$

The conjugate exponent $p = q = 2$ is special throughout the function spaces based on the Lebesgue integral. One can show, for example, that the only Hilbert space among the Hardy spaces is H^2, and a similar statement can be made about the Sobolev spaces. These spaces have found widespread use and have many special properties not enjoyed by the other L^p (H^p, $W^{k,p}$) spaces, $p \neq 2$, because they are Hilbert spaces. In the next section we investigate some special properties of Hilbert spaces.

For the remainder of this chapter we shall denote an inner product space $(X, \langle\cdot,\cdot\rangle)$ simply by X, unless there is some danger of confusion, and the norm $\|\cdot\|$ on X will always be the norm induced by the inner product unless otherwise stipulated.

Exercises 9-1:

1. Verify that the function $\langle\cdot,\cdot\rangle$ defined in Example 9-1-2 satisfies the conditions for an inner product.
2. Let X be an inner product space and suppose $f, g \in X$ with $g \neq 0$. Prove that $|\langle f, g/\|g\|\rangle| \leq \|f\|$ and use this to prove the Schwarz inequality.
3. Let X be an inner product space. Prove that for any elements $f, g, h \in X$,
$$\|h - f\|^2 + \|h - g\|^2 = \frac{1}{2}\|f - g\|^2 + 2\|h - \frac{1}{2}(f + g)\|^2.$$
This relation is known as the *Appolonius identity.*

9.2 Orthogonal Sets

The paradigm for a finite-dimensional inner product space is the space $\mathbb{R}^n$ discussed in Example 9-1-1. In this space, the inner product can be used to measure the angle between two vectors, i.e., $\langle \mathbf{x}, \mathbf{y}\rangle = \|\mathbf{x}\|\,\|\mathbf{y}\|\cos\phi$, where ϕ is the angle between the vectors $\mathbf{x}$ and $\mathbf{y}$. This geometrical idea can be extended to infinite-dimensional real inner product spaces, but there is no satisfactory extension to general inner product spaces. It turns out, however, that the magnitude of the angle between elements in an infinite-dimensional real vector space is of limited interest in the general theory with one important exception. Recall that for $\mathbf{x}, \mathbf{y} \in \mathbb{R}^n$ the relationship $\langle \mathbf{x}, \mathbf{y}\rangle = 0$ means geometrically that the vector $\mathbf{x}$ is orthogonal to the vector $\mathbf{y}$. This concept of orthogonality can be readily extended to general inner product spaces. Let $(X, \langle\cdot,\cdot\rangle)$ be any inner product space. Two elements $f, g \in X$ are said to be **orthogonal** if $\langle f, g\rangle = 0$.

This relationship is denoted by $f \perp g$. As we shall see, orthogonality plays an important part in Hilbert space theory.

Example 9-2-1:
Consider the inner product space $L^2[-\pi, \pi]$, and let f_n denote the functions defined by $f_n(x) = \sin(nx)$ for $n = 1, 2, \ldots$. Now,

$$\begin{aligned}\langle f_n, f_m \rangle &= \int_{[-\pi,\pi]} f_n(x) f_m(x)\,dx \\ &= \int_{-\pi}^{\pi} \sin(nx)\sin(mx)\,dx.\end{aligned}$$

Integration by parts indicates that

$$\begin{aligned}\langle f_n, f_m \rangle &= \left[-\frac{1}{m}\sin(nx)\cos(mx)\right]_{-\pi}^{\pi} + \frac{n}{m}\int_{-\pi}^{\pi} \cos(nx)\cos(mx)\,dx \\ &= \frac{n}{m}\int_{-\pi}^{\pi} \cos(nx)\cos(mx)\,dx,\end{aligned}$$

and integration again by parts yields

$$\begin{aligned}\langle f_n, f_m \rangle &= \frac{n}{m}\left\{\left[\frac{1}{m}\sin(mx)\cos(nx)\right]_{-\pi}^{\pi} + \frac{n}{m}\int_{-\pi}^{\pi} \sin(nx)\sin(mx)\,dx\right\} \\ &= \left(\frac{n}{m}\right)^2 \langle f_n, f_m \rangle.\end{aligned}$$

If $n \neq m$, then $1 - (n/m)^2 \neq 0$, and consequently $\langle f_n, f_m \rangle = 0$. If $n = m$, then

$$\langle f_n, f_m \rangle = \|f_n\|^2 = \int_{-\pi}^{\pi} \sin^2(nx)\,dx = \pi.$$

We therefore have that $f_n \perp f_m$ unless $m = n$.

Suppose that f and g are elements in the inner product space X. Then

$$\|f + g\|^2 = \langle f + g, f + g \rangle = \|f\|^2 + 2\mathrm{Re}\,\langle f, g \rangle + \|g\|^2.$$

If $f \perp g$ then $\langle f, g \rangle = 0$, and we thus have an extension of **Pythagoras's theorem**:

$$\|f + g\|^2 = \|f\|^2 + \|g\|^2.$$

We pause here to introduce two terms applicable to a general vector space. Let X be a vector space, and let Y and Z be subspaces of X. If for each $f \in X$ there exist elements $g \in Y$ and $h \in Z$ such that $f = g + h$, then we say that X is the **vector sum** of Y and Z, and denote this relationship by $X = Y + Z$. If, in addition, the elements $g \in Y$ and $h \in Z$ are determined *uniquely* for every $f \in X$, then we say that X is the **direct sum** of Y and Z, and write $X = Y \oplus Z$.

Given a set $S \subseteq X$, where $(X, \langle \cdot, \cdot \rangle)$ is an inner product space, another set $S^\perp$ can be formed by taking all the elements of X that are orthogonal to *every* element of S, i.e., the set $S^\perp = \{f \in X : \langle f, g \rangle = 0 \text{ for all } g \in S\}$. The set $S^\perp$ is called the **orthogonal complement** of S.

Example 9-2-2:
Let $\mathbf{e}_1, \mathbf{e}_2, \mathbf{e}_3$ be three linearly independent vectors in $\mathbb{R}^3$ such that $\langle \mathbf{e}_j, \mathbf{e}_k \rangle = 0$ for $j, k = 1, 2, 3$ unless $j = k$. If $S = \{\mathbf{v} \in \mathbb{R}^3 : \mathbf{v} = a\mathbf{e}_1 + b\mathbf{e}_2$ for some $a, b \in \mathbb{R}\}$ (i.e., the span of $\{\mathbf{e}_1, \mathbf{e}_2\}$), then $S^\perp = \{\mathbf{w} \in \mathbb{R}^3 : \mathbf{w} = c\mathbf{e}_3$ for some $c \in \mathbb{R}\}$. Since the set $\{\mathbf{e}_1, \mathbf{e}_2, \mathbf{e}_3\}$ forms a basis for $\mathbb{R}^3$, we have that $\mathbb{R}^3 = S \oplus S^\perp$.

Example 9-2-3:
Let ℓ^2 be the inner product space defined in Example 9-1-2, and let $\mathbf{e}_1 \in \ell^2$ be the sequence $\{1, 0, 0, \ldots\}$. Let S be the subspace defined by $S = \{\mathbf{a} \in \ell^2 : \mathbf{a} = \alpha\mathbf{e}_1 \text{ for some } \alpha \in \mathbb{C}\}$, and define the set B by $B = \{\mathbf{b} = \{b_n\} \in \ell^2 : b_1 = 0\}$. Then $\langle \mathbf{a}, \mathbf{b} \rangle = 0$ for all $\mathbf{a} \in S$ and all $\mathbf{b} \in B$, and consequently $B \subseteq S^\perp$. In fact, $B = S^\perp$, for if there is a sequence $\mathbf{c} = \{c_n\}$ with $c_1 \neq 0$, then $\langle \mathbf{a}, \mathbf{c} \rangle = \alpha c_1$, and αc_1 is not zero for *all* $\alpha \in \mathbb{C}$. Since any sequence $\mathbf{d} = \{d_n\}$ can be written as $\{d_1, 0, 0, \ldots\} + \{0, d_2, d_3, \ldots\}$, and this representation is unique, we have that $\ell^2 = S \oplus S^\perp$.

The inner product space $\mathbb{R}^3$ and the set S of Example 9-2-2 demonstrate two geometrical properties both of which extend to any finite dimensional inner product space. First, given any vector $\mathbf{x} \in \mathbb{R}^3$ there is a *unique* vector $\mathbf{v} \in S$ at a minimum distance from $\mathbf{x}$, i.e., $\|\mathbf{x} - \mathbf{v}\| < \|\mathbf{x} - \hat{\mathbf{v}}\|$ for all $\hat{\mathbf{v}} \in S \setminus \{v\}$. Second, any vector $\mathbf{x} \in \mathbb{R}^3$ can be represented in the form $\mathbf{x} = \mathbf{v} + \mathbf{w}$, where $\mathbf{v} \in S$ and $\mathbf{w} \in S^\perp$ are uniquely determined, i.e., $R^3 = S \oplus S^\perp$. Any subspace of $\mathbb{R}^3$ has these properties, and one can enquire whether these properties persist in the infinite-dimensional case. The space ℓ^2 in Example 9-2-3 demonstrates the second property, and it is straightforward to show that the set S in this example also has the "minimum distance" property. Unfortunately, these properties do not carry over to the general inner product space. Under the crucial assumption of completeness, however, these properties do extend to infinite-dimensional inner product spaces.

Theorem 9.2.1
Let S be a closed subspace of the Hilbert space X and let $f \in X$. Then there exists a unique $g \in S$ at a minimum distance from f.

Note that if S is a closed subspace, then S is a Hilbert space in its own right.

Theorem 9.2.2 (Projection Theorem)
Let S be a closed subspace of the Hilbert space X. Then $S^{\perp}$ is also a closed subspace in X, and $X = S \oplus S^{\perp}$. Moreover, if $f \in X$ is decomposed into the form $f = g + h$, where $g \in S$ and $h \in S^{\perp}$, then g is the unique element in S closest to f.

The projection theorem provides a key to understanding bases in Hilbert spaces. The study of bases in general Banach spaces is of limited value, but there is a rich and useful theory for bases in Hilbert spaces. As our main focus will eventually be on the space L^2, which is separable, we limit our general discussion to bases for separable Hilbert spaces.

In Example 9-2-2, consider the set $K = \{\mathbf{e}_1, \mathbf{e}_2, \mathbf{e}_3\}$. The set $K^{\perp}$ consists of all the vectors $\mathbf{x} \in \mathbb{R}^3$ such that $\langle \mathbf{x}, \mathbf{e}_k \rangle = 0$ for $k = 1, 2, 3$, but the only vector with this property is $\mathbf{x} = \mathbf{0}$; thus $K^{\perp} = \mathbf{0}$. Now, the set K forms a basis for $\mathbb{R}^3$, and this is characterized by the condition $K^{\perp} = \mathbf{0}$. Motivated by this observation, we can extend the concept of an orthogonal basis to general (separable) Hilbert spaces. Let $\mathcal{N}$ be a set of elements in the Hilbert space X. If $\mathcal{N}^{\perp} = \mathbf{0}$, then $\mathcal{N}$ is called a **total set**[1] in X. If $\mathcal{M}$ is a countable set in X and $\langle \mathbf{a}, \mathbf{b} \rangle = 0$ for all $\mathbf{a}, \mathbf{b} \in \mathcal{M}$, $\mathbf{a} \neq \mathbf{b}$, then $\mathcal{M}$ is called an **orthogonal set** in X. If, in addition, $\|\mathbf{a}\| = 1$ for all $\mathbf{a} \in \mathcal{M}$, then $\mathcal{M}$ is called an **orthonormal**

[1]These sets are also called "complete" in the literature. We avoid this term, since "complete" in this context has no connection with "complete" as used in Banach space theory.

set in X. For our purposes, we are interested primarily in the case where $\mathcal{M}$ has an infinite number of elements but is countable.

Let $\mathcal{M} = \{\phi_n\}$ be an orthonormal set in the Hilbert space X, and let $f \in X$. The numbers $\langle f, \phi_n\rangle$ are called the **Fourier coefficients** of f with respect to $\mathcal{M}$, and the series $\sum_{n=1}^{\infty}\langle f, \phi_n\rangle\phi_n$ is called the **Fourier series** of f (with respect to $\mathcal{M}$). Now, for any $m \in \mathbb{N}$,

$$\begin{aligned}
0 \le \|f - \sum_{n=1}^{m}\langle f, \phi_n\rangle\phi_n\|^2 &= \langle f - \sum_{n=1}^{m}\langle f, \phi_n\rangle\phi_n, f - \sum_{n=1}^{m}\langle f, \phi_n\rangle\phi_n\rangle \\
&= \langle f, f\rangle - \langle f, \sum_{n=1}^{m}\langle f, \phi_n\rangle\phi_n\rangle - \langle\sum_{n=1}^{m}\langle f, \phi_n\rangle\phi_n, f\rangle \\
&\quad + \langle\sum_{n=1}^{m}\langle f, \phi_n\rangle\phi_n, \sum_{n=1}^{m}\langle f, \phi_n\rangle\phi_n\rangle \\
&= \|f\|^2 - 2\sum_{n=1}^{m}\langle f, \phi_n\rangle\overline{\langle f, \phi_n\rangle} + \|\sum_{n=1}^{m}\langle f, \phi_n\rangle\phi_n\|^2 \\
&= \|f\|^2 - 2\sum_{n=1}^{m}|\langle f, \phi_n\rangle|^2 + \|\sum_{n=1}^{m}\langle f, \phi_n\rangle\phi_n\|^2,
\end{aligned}$$

and Pythagoras's equation implies that

$$\begin{aligned}
\|\sum_{n=1}^{m}\langle f, \phi_n\rangle\phi_n\|^2 &= \sum_{n=1}^{m}\|\langle f, \phi_n\rangle\phi_n\|^2 \\
&= \sum_{n=1}^{m}|\langle f, \phi_n\rangle|^2\|\phi_n\|^2 \\
&= \sum_{n=1}^{m}|\langle f, \phi_n\rangle|^2,
\end{aligned}$$

and therefore we have

$$\begin{aligned}
0 &\le \|f - \sum_{n=1}^{m}\langle f, \phi_n\rangle\phi_n\|^2 \\
&= \|f\|^2 - \sum_{n=1}^{m}|\langle f, \phi_n\rangle|^2.
\end{aligned}$$

Hence,

$$\sum_{n=1}^{m} |\langle f, \phi_n \rangle|^2 \leq \|f\|^2,$$

and since the sequence $\{\sigma_m\} = \{\sum_{n=1}^{m} |\langle f, \phi_n \rangle|^2\}$ is a monotonic increasing sequence bounded above, we have that $\{\sigma_m\}$ converges to some $\sigma \in \mathbb{R}$ and **Bessel's inequality** holds:

$$\sum_{n=1}^{\infty} |\langle f, \phi_n \rangle|^2 \leq \|f\|^2.$$

Bessel's inequality indicates that the Fourier series of f with respect to $\mathcal{M}$ is absolutely convergent (in the norm). This means that the Fourier series is convergent (in the norm) and that its sum is independent of the order in which terms are added. Thus, the series $\sum_{n=1}^{\infty} \langle f, \phi_n \rangle \phi_n$ converges to some function $P_{\mathcal{M}} f \in X$. Now, if $f \in [\mathcal{M}]^{\perp}$ $(= \mathcal{M}^{\perp})$, all the Fourier coefficients of f are zero, and thus $P_{\mathcal{M}} f = \mathbf{0}$. Alternatively, if $f \in \overline{[\mathcal{M}]}$, then it can be shown that $P_{\mathcal{M}} f = f$. For the general $f \in X$, the set $\overline{[\mathcal{M}]}$ is a closed subspace in the Hilbert space X, and the projection theorem indicates that $f = P_{\mathcal{M}} f + h$, where $P_{\mathcal{M}} f \in \overline{[\mathcal{M}]}$ and $h \in [\mathcal{M}]^{\perp}$. The function $P_{\mathcal{M}} f$ is thus the "closest" approximation in $\overline{[\mathcal{M}]}$ to f.

An orthonormal set $\mathcal{M} = \{\phi_n\}$ in the separable Hilbert space X is called an **orthonormal basis** of X if for every $f \in X$,

$$f = \sum_{n=1}^{\infty} \langle f, \phi_n \rangle \phi_n.$$

If $\mathcal{M}$ is an orthonormal basis, then

$$\lim_{m \to \infty} \sum_{n=1}^{m} \langle f, \phi_n \rangle \phi_n = f,$$

and using the derivation of Bessel's inequality (and continuity of the norm function) we have that

$$\begin{aligned} \lim_{m \to \infty} \left(\|f\|^2 - \sum_{n=1}^{m} |\langle f, \phi_n \rangle|^2 \right) &= \lim_{m \to \infty} \|f - \sum_{n=1}^{m} \langle f, \phi_n \rangle \phi_n\|^2 \\ &= 0. \end{aligned}$$

We thus arrive at **Parseval's formula**

$$\|f\|^2 = \sum_{n=1}^{\infty} |\langle f, \phi_n \rangle|^2.$$

Suppose now that $f \in \mathcal{M}^{\perp}$. Since $\mathcal{M}$ is an orthonormal basis of X, $f = \sum_{n=1}^{\infty} \langle f, \phi_n \rangle \phi_n$ and $\langle f, \phi_n \rangle = 0$ for all $\phi_n \in \mathcal{M}$; consequently, by Parseval's formula we have that $\|f\| = 0$, and the definition of a norm implies that $f = \mathbf{0}$. In this manner we see that if $\mathcal{M}$ is an orthonormal basis, then $\mathcal{M}^{\perp} = \mathbf{0}$, i.e., $\mathcal{M}$ is a total orthonormal set.

Yet another implication of $\mathcal{M}$ being an orthonormal basis is that $\overline{[\mathcal{M}]} = X$. This follows from the projection theorem, since $\overline{[\mathcal{M}]}^{\perp} \subseteq \mathcal{M}^{\perp} = \mathbf{0}$, and so

$$X = \overline{[\mathcal{M}]} \oplus \overline{[\mathcal{M}]}^{\perp} = \overline{[\mathcal{M}]} \oplus \mathbf{0} = \overline{[\mathcal{M}]}.$$

It is interesting that the above implications actually work the other way as well. For example, if $\mathcal{M}$ is an orthonormal set in X such that Parseval's formula is satisfied for all $f \in X$, then it can be shown that $\mathcal{M}$ is an orthonormal basis. In summary we have the following result:

Theorem 9.2.3
Let X be a separable Hilbert space and suppose that $\mathcal{M} = \{\phi_n\}$ is an orthonormal set in X. Then the following conditions are equivalent:

(i) *$\mathcal{M}$ is an orthonormal basis;*
(ii) *$\|f\|^2 = \sum_{n=1}^{\infty} |\langle f, \phi_n \rangle|^2$ for all $f \in X$;*
(iii) *$\mathcal{M}$ is a total set;*
(iv) *$\overline{[\mathcal{M}]} = X$.*

It can be shown that every separable Hilbert space has an orthonormal basis. Formally, we have the following result:

Theorem 9.2.4
Let X be a separable Hilbert space. Then there exists an orthonormal basis for X.

The proof of this result is based on a Gram–Schmidt process analogous to that used in linear algebra to derive orthonormal bases. Parseval's formula is remarkable in this context because it essentially identifies all separable Hilbert spaces with the space ℓ^2. Now, ℓ^2 is a separable Hilbert space, and given an orthonormal basis $\mathcal{M} = \{\phi_n\}$

on a general separable Hilbert space X there is a mapping $T : X \to \ell^2$ defined by the Fourier coefficients $\langle f, \phi_n \rangle$. In other words, T maps $f \in X$ to the sequence $\{\langle f, \phi_n \rangle\} \in \ell^2$. Parseval's formula shows that T is an *isometry* from X to ℓ^2. On the other hand, given any sequence $\{\alpha_n\} \in \ell^2$, the function defined by $\sum_{n=1}^{\infty} \alpha_n \phi_n$ is in X, since X is a Hilbert space and the series is convergent. We thus have that *any* separable Hilbert space is isometric to the Hilbert space ℓ^2. In fact, an even stronger result is available, viz., all separable complex Hilbert spaces are *isomorphic* to ℓ^2. This means that at an "algebraic level" a complex separable Hilbert space is indistinguishable from the space ℓ^2, i.e., at this level there is only one distinct space. A similar statement is true for real Hilbert spaces, where the space ℓ^2 is replaced by its real analogue.

In passing we note that any Hilbert space X possesses an "orthogonal basis," though it may not be countable. The cardinality of the set is called the *Hilbert dimension* of the space. If X and $\hat{X}$ are two Hilbert spaces both real or both complex with the same Hilbert dimension, then it can be shown that X and $\hat{X}$ are isomorphic. This is a generalization of the situation in finite-dimensional spaces, where, for example, all n-dimensional real Hilbert spaces are isomorphic to $\mathbb{R}^n$.

Exercises 9-2:

1. Show that for any integers m, n

$$\int_{[-\pi,\pi]} \cos(mx)\cos(nx)\,dx = \begin{cases} 0, & \text{if } m \neq n, \\ \pi, & \text{if } m = n, \end{cases}$$

   and

$$\int_{[-\pi,\pi]} \sin(mx)\cos(nx)\,dx = 0.$$

2. The first three *Legendre polynomials* are defined by $P_0(x) = 1$, $P_1(x) = x$, and $P_2(x) = \frac{1}{2}(3x^2 - 1)$.
   (a) Show that the set $P = \{P_0, P_1, P_2\}$ forms an orthogonal set on $L^2[-1, 1]$.
   (b) Construct an orthonormal set $\mathcal{M}$ from the set P and find the Fourier coefficients for the function $f(x) = e^x$.

3. The *Rademacher functions* are defined by

$$r_n(x) = \operatorname{sgn}(\sin(2^n \pi x))$$

for $n = 1, 2, \ldots$, where sgn denotes the *signum function*

$$\operatorname{sgn}(x) = \begin{cases} -1, & \text{if } x < 0, \\ 1, & \text{if } x > 0. \end{cases}$$

(a) Show that the set $\mathcal{M} = \{r_n\}$ is an orthogonal set on $L^2[0,1]$.

(b) If f is defined by $f(x) = \cos(2\pi x)$, show that $f \perp r_n$ for all $n = 1, 2, \ldots$ and deduce that $\mathcal{M}$ cannot form a total orthogonal set.

4. Let X be an inner product space and let $\mathcal{M} = \{\phi_n\}$ be an orthonormal basis of X. Show that for any $f, g \in X$,

$$\langle f, g\rangle = \sum_{k=1}^{\infty} \langle f, \phi_k\rangle \overline{\langle g, \phi_k\rangle}.$$

9.3 Classical Fourier Series

We saw in Section 9-1 that L^2 is a Hilbert space, and we know from Theorem 8.6.1 that all L^p spaces are separable. Theorem 9.2.4 thus implies that the space L^2 must have an orthonormal basis. It turns out that the classical Fourier series (i.e., trigonometric series) lead to an orthonormal basis for $L^2[a,b]$. In this section we study classical Fourier series and present some basic results with little detail. There are many specialized texts on the subject of Fourier series such as [15] and [45], and we refer the reader to these works for most the details. A particularly lively account of the theory, history, and applications of Fourier series can be found in [24]. For convenience we focus primarily on the space $L^2[-\pi, \pi]$ and note here that the results can be extended *mutatis mutandis* to the general closed interval.

A classical Fourier series is a series of the form

$$\frac{1}{2}a_0 + \sum_{n=1}^{\infty} \left(a_n \cos(nx) + b_n \sin(nx)\right), \tag{9.1}$$

where the a_n's and b_n's are constants. We know from Example 9-2-1 that for any nonzero integers m, n,

$$\int_{[-\pi,\pi]} \sin(mx)\sin(nx)\,dx = \begin{cases} 0, & \text{if } m \neq n, \\ \pi, & \text{if } m = n. \end{cases}$$

We also know from Exercises 9-2, No. 1, that for all integers m, n,

$$\int_{[-\pi,\pi]} \cos(mx)\cos(nx)\,dx = \begin{cases} 0, & \text{if } m \neq n, \\ \pi, & \text{if } m = n, \end{cases}$$

and

$$\int_{[-\pi,\pi]} \sin(mx)\cos(nx)\,dx = 0.$$

In addition, it is evident that for any integer n,

$$\int_{[-\pi,\pi]} \sin(nx)\,dx = \int_{[-\pi,\pi]} \cos(nx)\,dx = 0.$$

Let $\Phi = \{\phi_n\}$ and $\Psi = \{\psi_n\}$ denote the sets of functions defined by

$$\phi_n(x) = \frac{\cos(nx)}{\sqrt{\pi}}, \quad \psi_n(x) = \frac{\sin(nx)}{\sqrt{\pi}},$$

for $n = 1, 2, \ldots$. Then the above relationships indicate that the set $S = \{1/\sqrt{2\pi}\} \cup \Phi \cup \Psi$ forms an orthonormal set in $L^2[-\pi,\pi]$. The Fourier coefficients of a function $f \in L^2[-\pi,\pi]$ with respect to S are given by

$$\begin{aligned} \left\langle f, \frac{1}{\sqrt{2\pi}} \right\rangle &= \frac{1}{\sqrt{2\pi}} \int_{[-\pi,\pi]} f(x)\,dx, \\ \langle f, \phi_n \rangle &= \frac{1}{\sqrt{\pi}} \int_{[-\pi,\pi]} f(x)\cos(nx)\,dx, \\ \langle f, \psi_n \rangle &= \frac{1}{\sqrt{\pi}} \int_{[-\pi,\pi]} f(x)\sin(nx)\,dx, \end{aligned}$$

and the Fourier series is

$$P_S f = \left\langle f, \frac{1}{\sqrt{2\pi}} \right\rangle \frac{1}{\sqrt{2\pi}} + \sum_{n=1}^{\infty} \left(\langle f, \phi_n \rangle \phi_n + \langle f, \psi_n \rangle \psi_n \right), \tag{9.2}$$

which is equivalent to expression (9.1) with the familiar coefficient relations

$$a_0 = \frac{1}{\pi}\int_{[-\pi,\pi]} f(x)\,dx,$$

$$a_n = \frac{1}{\pi}\int_{[-\pi,\pi]} f(x)\cos(nx)\,dx,$$

$$b_n = \frac{1}{\pi}\int_{[-\pi,\pi]} f(x)\sin(nx)\,dx.$$

Fourier series can be expressed in a tidier, more symmetric, form using the relation $e^{inx} = \cos(nx) + i\sin(nx)$. Using this relation, the series (9.1) can be written in the form

$$\sum_{-\infty}^{\infty} c_n e^{inx}, \tag{9.3}$$

where the c_n are complex numbers defined by

$$c_n = \frac{1}{2\pi}\int_{[-\pi,\pi]} f(x)e^{inx}\,dx.$$

The set of functions $B = \{\beta_n\}$, where

$$\beta_n(x) = \frac{e^{inx}}{\sqrt{2\pi}},$$

forms an orthonormal set for $L^2_{\mathbb{C}}[-\pi,\pi]$, and for any $f \in L^2_{\mathbb{C}}[-\pi,\pi]$ (and consequently for any $f \in L^2[-\pi,\pi]$) the corresponding Fourier series is

$$P_B f = \sum_{-\infty}^{\infty} \langle f, \beta_n\rangle \beta_n. \tag{9.4}$$

Now, for any $f \in L^2[-\pi,\pi]$, Bessel's inequality guarantees that the Fourier series (9.2) (and (9.3)) converges in the $\|\cdot\|_2$ norm to some function $P_S f \in L^2[-\pi,\pi]$. The central question here is whether the set S forms an orthonormal basis for $L^2[-\pi,\pi]$, so that $P_S f = f$ a.e. In fact, it can be shown that S forms an orthonormal basis for $L^2[-\pi,\pi]$. If we combine this fact with Theorem 9.2.3, we have the following result:

Theorem 9.3.1
Let $f \in L^2[-\pi, \pi]$ and let

$$s_n = \left\langle f, \frac{1}{\sqrt{2\pi}} \right\rangle \frac{1}{\sqrt{2\pi}} + \sum_{k=1}^{n} (\langle f, \phi_k \rangle \phi_k + \langle f, \psi_k \rangle \psi_k)$$
$$= \sum_{k=-n}^{k=n} \langle f, \beta_k \rangle \beta_k.$$

Then:

(i) $\|s_n - f\|_2 \to 0$ *as* $n \to \infty$;
(ii) *Parseval's relation is satisfied:*

$$\|f\|_2^2 = |\langle f, 1 \rangle|^2 + \sum_{n=1}^{\infty} \left(|\langle f, \phi_n \rangle|^2 + |\langle f, \psi_k \rangle|^2 \right)$$
$$= \frac{\pi}{2} a_0^2 + \pi \sum_{n=1}^{\infty} (a_n^2 + b_n^2)$$
$$= \sum_{-\infty}^{\infty} |\langle f, \beta_n \rangle|^2 = 2\pi \sum_{-\infty}^{\infty} |c_n|^2.$$

From the last section we know that all separable Hilbert spaces are isomorphic to the Hilbert space ℓ^2, and Parseval's relation is a manifestation of this relationship. This observation leads to the following result:

Theorem 9.3.2 (Riesz–Fischer)
Let $\{c_n\} \in \ell^2$. Then there is a unique function $f \in L^2[-\pi, \pi]$ such that the c_n are the Fourier coefficients for f.
Note that "unique" in the above theorem means that the sequence $\{c_n\}$ determines an equivalence class of functions modulo equality a.e.

An immediate consequence of Theorem 9.3.1 and the projection theorem (Theorem 9.2.2) is the following result:

Theorem 9.3.3
Let $f \in L^2[-\pi, \pi]$. Then for any $\epsilon > 0$ there exists a positive integer n and a trigonometric polynomial of degree n, say $Q_n = \sum_{k=-n}^{n} q_k \beta_k$ such that $\|Q_n - f\|_2 < \epsilon$. Moreover, among the trigonometric polynomials of

degree n, the closest approximation to f in the $\|\cdot\|_2$ norm is that for which the q_k correspond to the Fourier coefficients.

In other words, trigonometric polynomials can be used to approximate any function in $L^2[-\pi,\pi]$, and the Fourier coefficients provide the best approximation among such polynomials in the $\|\cdot\|_2$ norm.

Theorem 9.3.1 guarantees that the Fourier series will converge in the $\|\cdot\|_2$ norm, but this is not the same as pointwise convergence, and an immediate question is whether or not a Fourier series for a function $f \in L^2[-\pi,\pi]$ converges pointwise to f. More explicitly, for a fixed $x \in [-\pi,\pi]$ does the sequence of numbers $\{s_n(x)\}$ converge, and if so does $s_n(x) \to f(x)$ as $n \to \infty$? The answer to this question is complicated, and much of the research on Fourier series revolved around pointwise convergence. Some simple observations can be made. First, the orthonormal basis defining the Fourier series is manifestly periodic, so that if $s_n(x)$ converges for $x = -\pi$, it also converges for $x = \pi$, and $s(-\pi) = s(\pi)$. The existence of a Fourier series does not require that $f(-\pi) = f(\pi)$, so pointwise convergence will fail at an endpoint unless f satisfies this condition. More generally, we see that any two integrable functions f and g such that $f = g$ a.e. produce the same Fourier coefficients, so that for a specific function f, we expect that the best generic situation would be that $s_n(x) \to f(x)$ a.e. We shall not "plumb the depths" of the vast results concerning pointwise convergence of Fourier series; however, we will discuss two results of interest in their own right that make crucial use of the Lebesgue integral.

Prima facie, it is not obvious that the coefficients a_n and b_n in the Fourier series have limit zero as $n \to \infty$. Given that arguments such as $x = 0$ lead to series such as $\sum_{n=0}^{\infty} a_n$, this is clearly a concern. An elegant result called the Riemann–Lebesgue theorem resolves this concern and is of interest in its own right. We state the result here in a form more general than is required for the question at hand.

Theorem 9.3.4 (Riemann–Lebesgue)
Let $I \subseteq \mathbb{R}$ be some interval and $f \in L^1(I)$. If $\{\lambda_n\}$ is a sequence of real numbers such that $\lambda_n \to \infty$ as $n \to \infty$, then

$$\int_I f(x)\cos(\lambda_n x)\,dx \to 0 \quad \text{and} \quad \int_I f(x)\sin(\lambda_n x)\,dx \to 0$$

as $n \to \infty$.

Proof The proof of this result is of particular interest because the convergence theorems of Chapter 5 are of little help, and we must return to the definition of the integral itself. We sketch here the proof for the cosine integral.

Suppose first that $I = [a, b]$ and that $f : I \to \mathbb{R}$ is bounded on I. If M denotes an upper bound for $|f|$, then

$$\begin{aligned} \left|\int_{[a,b]} f(x)\cos(\lambda_n x)\,dx\right| &\le M\left|\int_{[a,b]} \cos(\lambda_n x)\,dx\right| \\ &= M\left|\int_a^b \cos(\lambda_n x)\,dx\right| \\ &= \frac{M}{\lambda_n}\left|\sin(\lambda_n b) - \sin(\lambda_n a)\right| \\ &\le \frac{2M}{\lambda_n}. \end{aligned}$$

Since $\lambda_n \to \infty$, as $n \to \infty$, we have that $\int_{[a,b]} f(x)\cos(\lambda_n x)\,dx \to 0$. This calculation shows in particular that the theorem holds for any step function having a support of finite length. Now, the Lebesgue integral is defined in terms of a sequence of α-summable functions, with $\alpha = x$, and by definition these functions have supports of finite length. If $f \in L^1(I)$, then we *know* that there is a sequence θ_j of x-summable functions such that

$$\int_I |f(x) - \theta_j(x)|\,dx \to 0$$

as $j \to \infty$. Choose any $\epsilon > 0$, and select a j sufficiently large so that

$$\int_I |f(x) - \theta_j(x)|\,dx < \frac{\epsilon}{2}.$$

Now,

$$\begin{aligned} &\left|\int_I f(x)\cos(\lambda_n x)\,dx\right| \\ &\quad = \left|\int_I (f(x) - \theta_j(x) + \theta_j(x))\cos(\lambda_n x)\,dx\right| \\ &\quad \le \left|\int_I (f(x) - \theta_j(x))\cos(\lambda_n x)\,dx\right| + \left|\int_I \theta_j(x)\cos(\lambda_n x)\,dx\right| \end{aligned}$$

$$\leq \int_I |f(x) - \theta_j(x)|\, dx + \left| \int_I \theta_j(x) \cos(\lambda_n x)\, dx \right|$$
$$< \frac{\epsilon}{2} + \left| \int_I \theta_j(x) \cos(\lambda_n x)\, dx \right|.$$

From the above discussion we know that

$$\int_I \theta_j(x) \cos(\lambda_n x)\, dx \to 0$$

as $n \to \infty$ for any j, so that there is an integer N such that

$$\left| \int_I \theta_j(x) \cos(\lambda_n x)\, dx \right| < \frac{\epsilon}{2}$$

whenever $n > N$. Therefore,

$$\left| \int_I f(x) \cos(\lambda_n x)\, dx \right| < \frac{\epsilon}{2} + \frac{\epsilon}{2} = \epsilon$$

whenever $n > N$, and by definition this means that

$$\int_I f(x) \cos(\lambda_n x)\, dx \to 0$$

as $n \to \infty$. □

The Riemann–Lebesgue theorem with $I = [-\pi, \pi]$ and $\lambda_n = n$ shows that the Fourier coefficients tend to zero as $n \to \infty$. This theorem is also crucial in proving another notable result known as the Riemann localization theorem, which we shall not prove.

Theorem 9.3.5 (Riemann Localization)
Let $f \in L^1[-\pi, \pi]$ and $x \in [-\pi, \pi]$. Then for any fixed δ with $0 < \delta < \pi$, $s_n(x) \to f(x)$ if and only if

$$\lim_{n\to\infty} \frac{1}{2\pi} \int_{[0,\delta]} \frac{f(x+t) + f(x-t)}{t} \sin((n + \frac{1}{2})t)\, dt = 0.$$

Here it is assumed that f has been extended periodically to a function on $\mathbb{R}$ so that f is defined for arguments $x \pm t$ that may not be in $[-\pi, \pi]$. What is interesting in the above theorem is that the number δ can be *arbitrarily small*, and this means that the pointwise convergence of the Fourier series depends only on the values that f

assumes in a small neighborhood of x. Given that the Fourier coefficients in the series defining s_n depend on the values f assumes in the entire interval $[-\pi, \pi]$, this result is remarkable.

We can study Fourier series outside the comfortable space $L^2[-\pi, \pi]$. Naturally, we lose the results that rely on $L^2[-\pi, \pi]$ being a Hilbert space such as Parseval's relation, but in the larger space $L^1[-\pi, \pi]$ the Fourier coefficients are still well-defined, and results such as the Riemann–Lebesgue theorem are still valid. We even have the following uniqueness result:

Theorem 9.3.6
If $f, g \in L^1[-\pi, \pi]$ have the same Fourier coefficients, then $f = g$ a.e.
What is needed, however, is some result that shows that the partial sums of the Fourier series for a function $f \in L^1[-\pi, \pi]$ converge in the $\|\cdot\|_1$ norm to f, but this is where things go wrong. If $f \in L^1[-\pi, \pi]$, then in general we do not have that $\|s_n - f\|_1 \to 0$ as $n \to \infty$. It can be shown, however, that for

$$\sigma_n = \frac{1}{n} \sum_{k=1}^{n-1} s_k,$$

$\|\sigma_n - f\|_1 \to 0$ as $n \to \infty$ for any $f \in L^1[-\pi, \pi]$. The quantity σ_n is called the *Cesàro mean* of the partial sum sequence $\{s_n\}$. This is a weaker convergence result, since σ_n effectively measures an averaged partial sum. The sequence $\{\sigma_n\}$ may converge even if $\{s_n\}$ diverges, and if $\{s_n\}$ converges to some s, then $\{\sigma_n\}$ also converges to some function $s \in L^1[-\pi, \pi]$. In the space $L^1[-\pi, \pi]$ this is the sharpest result we can get. In fact, it can be shown that there are functions in $L^1[-\pi, \pi]$ such that $\{s_n(x)\}$ *diverges a.e.* Though most of the above results concerning pointwise convergence were established by the early twentieth century, it was not until the 1960's that Carleson [7] proved that is $f \in L^2[-\pi, \pi]$ then $\{s_n(x)\} \to f(x)$ a.e. From this perspective, the space $L^2[-\pi, \pi]$ is the natural space in which to study Fourier series.

Exercises 9-3:

1. Let $f(x) = x$ on the interval $[-\pi, \pi]$.

   (a) Determine the Fourier series for f.

(b) Show that the series obtained by differentiating term by term the Fourier series in part 1(a) is a divergent series.

2. The complex Fourier series for a function $f \in L^2[-A, A]$, $A > 0$, is given by

$$P_{\mathcal{M}} f(x) = \sum_{-\infty}^{\infty} c_n e^{in\pi x/A},$$

where

$$c_n = \frac{1}{2M} \int_{-A}^{A} f(x) e^{-in\pi x/A}\, dx,$$

for $n = 0, \pm 1, \pm 2, \ldots$, and it can be shown that the set $\mathcal{M} = \{e^{in\pi x/A}\}$ suitably normalized forms an orthonormal basis for $L^2[-A, A]$.

(a) Find the complex Fourier series for the function $g : \left[-\frac{1}{2}, \frac{1}{2}\right] \to \mathbb{R}$ defined by

$$g(x) = \begin{cases} -1, & \text{if } -\frac{1}{2} < x < 0, \\ 1, & \text{if } 0 < x < \frac{1}{2}. \end{cases}$$

(b) Use Parseval's relation and the Fourier series in 2(a) to prove that

$$\sum_{n=0}^{\infty} \frac{1}{(2n+1)^2} = \frac{\pi^2}{8}.$$

3. Suppose that f has a continuous derivative on the interval $[-\pi, \pi]$, and let a_n, b_n denote the Fourier coefficients of f. Use integration by parts to prove the Riemann–Lebesgue theorem for this special case.

9.4 The Sturm–Liouville Problem

Theorem 9.2.4 indicates that for any interval I the Hilbert space $L^2(I)$ must have an orthonormal basis. If I is bounded, then the trigonometric functions provide such a basis. Theorem 9.2.4 does not, however, preclude the existence of other orthonormal bases, and there are, in fact, any number of orthonormal bases available. If

I is not bounded, then the trigonometric functions are not even in the space $L^2(I)$, and the classical Fourier expansions are no longer valid. Nonetheless, Theorem 9.2.4 guarantees the existence of an orthonormal basis for any interval I.

In practice, many of the orthonormal bases for $L^2(I)$ arise as solutions to boundary value problems of the Sturm–Liouville type. A *regular Sturm–Liouville problem* consists in determining a solution y to a differential equation of the form

$$\frac{d}{dx}\left(r(x)\frac{dy}{dx}\right) + (q(x) + \lambda p(x))y = 0 \tag{9.5}$$

on some interval $[a, b]$, satisfying boundary conditions of the form

$$k_1 y(a) + k_2 y'(a) = 0, \;\; l_1 y(b) + l_2 y'(b) = 0. \tag{9.6}$$

Here, r, q, and p are given functions, λ is a complex parameter, and the k_j's and l_j's are constants such that $k_1^2 + k_2^2 \neq 0$, $l_1^2 + l_2^2 \neq 0$. In addition, it is assumed for the regular Sturm–Liouville problem that the functions r and p are nonzero in the interval $[a, b]$.

Sturm–Liouville problems are important in theory and applications of differential equations. As a consequence, this class of boundary value problems has been studied exhaustively for some 150 years. Relatively accessible accounts of the theory can be found in [5], [9], [11], [23], and [40]. In addition, many applications-oriented texts such as [3], [8], and [26] contain short summaries of the theory and various applications. We do not attempt here to replicate the general theory in any detail or depth, but merely focus on the Sturm–Liouville problem as a "machine" for producing orthogonal sets in L^2.

Equation (9.5) is often written in the abbreviated form

$$\mathcal{L}y = -\lambda p(x)y, \tag{9.7}$$

where the linear operator $\mathcal{L}$ is defined by

$$\mathcal{L}y = \frac{d}{dx}\left(r(x)\frac{dy}{dx}\right) + q(x)y. \tag{9.8}$$

If $y \in C^2[a, b]$, $r \in C'[a, b]$, and $q \in C[a, b]$ then $Ly \in C[a, b]$, and in particular $\mathcal{L}y \in L^2[a, b]$. For linear operators such as $\mathcal{L}$ there exists another operator $\mathcal{L}^*$ called the (Hilbert) adjoint. The adjoint operator satisfies the relation

$$\langle \mathcal{L}y_1, y_2 \rangle = \langle y_1, \mathcal{L}^* y_2 \rangle \tag{9.9}$$

for all $y_1, y_2 \in L^2[a,b]$, which also satisfy the boundary conditions (9.6). The peculiar form of the operator $\mathcal{L}$ for the Sturm–Liouville problem ensures that the operator is *self-adjoint*, i.e., $\mathcal{L} = \mathcal{L}^*$. This can be verified directly as follows:

$$\begin{aligned}
&\langle \mathcal{L}y_1, y_2 \rangle - \langle y_1, \mathcal{L}y_2 \rangle \\
&= \int_{[a,b]} \left(y_2(x)\mathcal{L}y_1(x) - y_1(x)\mathcal{L}y_2(x) \right) dx \\
&= \int_{[a,b]} \frac{d}{dx}\left(\left(r(x)\frac{d}{dx}y_1(x) \right) y_2(x) - \left(r(x)\frac{d}{dx}y_2(x) \right) y_1(x) \right) dx \\
&= \left[r(x) \left(y_2(x)\frac{d}{dx}y_1(x) - y_1(x)\frac{d}{dx}y_2(x) \right) \right]_a^b \\
&= 0.
\end{aligned}$$

The final equality follows from the condition that y_1 and y_2 satisfy the same boundary conditions (9.6). (The specific details can be found in [26]). As we shall see shortly, the self-adjointness of $\mathcal{L}$ is the key to obtaining orthogonal sets.

For any value $\lambda \in \mathbb{C}$, the Sturm–Liouville problem always has one obvious solution, viz., the trivial solution $y \equiv 0$. Generically, this is the only solution available for a given value of λ, but there may be some values of λ for which nontrivial solutions exist. These values are called **eigenvalues**, and the associated nontrivial solutions are called **eigenfunctions**. The set of all eigenvalues for the problem is called the **spectrum**[2] of the problem. The theory of the existence and distribution of eigenvalues to problems such as the Sturm–Liouville problem forms a significant component of linear functional analysis known as spectral analysis. Most introductory texts on functional analysis such as [25] devote a few chapters to the general theory. A comprehensive and detailed discussion of the subject can be found in [12].

---

[2] More generally, the spectrum is the set of points where the inverse operator (the resolvent) $(\mathcal{L} - \lambda I)^{-1}$ is not well-defined.

The regular Sturm–Liouville problem has fairly tractable spectral properties, as the next theorem illustrates.

Theorem 9.4.1

Suppose that $r \in C^1[a, b]$, $p, q \in C[a, b]$, and that $r(x) > 0$ and $p(x) > 0$ for all $x \in [a, b]$. Then:

(i) the spectrum for the Sturm–Liouville problem is an infinite but countable set;

(ii) the eigenvalues are all real;

(iii) to each eigenvalue there corresponds precisely one eigenfunction (up to a constant factor), i.e., the eigenvalues are simple;

(iv) the spectrum contains no finite accumulation points.

Thus, under the conditions of the above theorem, the Sturm–Liouville problem always has an infinite (but countable) number of eigenfunctions. The spectrum is a countable set, so we can regard it as a sequence $\{\lambda_n\}$ and impose the condition $|\lambda_n| \leq |\lambda_{n+1}|$ for all n. Since there are no finite accumulation points, we see that $|\lambda_n| \to \infty$ as $n \to \infty$.

Suppose that the conditions of Theorem 9.4.1 are satisfied, and that λ_m and λ_n are distinct eigenvalues with corresponding eigenfunctions y_m, y_n, respectively. Then,

$$\mathcal{L}y_m = -\lambda_m p(x) y_m,$$
$$\mathcal{L}y_n = -\lambda_n p(x) y_n,$$

and therefore

$$y_m \mathcal{L} y_n - y_n \mathcal{L} y_m = (\lambda_m - \lambda_n) p(x) y_m y_n.$$

The above equation indicates that

$$\begin{aligned}\int_{[a,b]} &\left(y_m(x)\mathcal{L}y_n(x) - y_n(x)\mathcal{L}y_m(x)\right) dx \\ &= \langle y_m, \mathcal{L}y_n\rangle - \langle y_n, \mathcal{L}y_m\rangle \\ &= (\lambda_m - \lambda_n)\int_{[a,b]} p(x) y_m(x) y_n(x)\, dx.\end{aligned}$$

Now, the eigenfunctions are real, so that $\langle y_m, \mathcal{L}y_n\rangle = \langle \mathcal{L}y_n, y_m\rangle$, and since $\mathcal{L}$ is self-adjoint,

$$(\lambda_m - \lambda_n)\int_{[a,b]} p(x) y_m(x) y_n(x)\, dx = \langle \mathcal{L}y_n, y_m\rangle - \langle y_n, \mathcal{L}y_m\rangle = 0.$$

By hypothesis, the eigenvalues are distinct ($\lambda_m \neq \lambda_n$), and the above calculation shows that

$$\int_{[a,b]} p(x) y_m(x) y_n(x)\, dx = 0. \tag{9.10}$$

If $p = 1$, the above equation implies that $y_m \perp y_n$ for $m \neq n$, and thus the eigenfunctions are orthogonal. By hypothesis we have that $p(x) > 0$, so that in any event the set of functions $\{\sqrt{p} y_n\}$ is orthogonal. As the eigenfunctions are by definition nontrivial solutions, we know that $\|py_n\|_2 \neq 0$, so that this set of functions can always be normalized to form an orthonormal set in $L^2[a, b]$. The Sturm–Liouville problem thus produces eigenfunctions from which orthonormal sets can be derived.

Given a continuous function p positive on the interval $[a, b]$, it is always possible to define another norm for $L^2[a, b]$, viz.,

$$_p\|y\|_2 = \left\{ \int_{[a,b]} p(x) y^2(x)\, dx \right\}^{1/2}. \tag{9.11}$$

Since p is positive and continuous on $[a, b]$, there exist numbers p_m and p_M such that $0 < p_m \leq p(x) \leq p_M$ for all $x \in [a, b]$. This implies that

$$p_m \|y\|_2 \leq {}_p\|y\|_2 \leq p_M \|y\|_2,$$

so that the $_p\|\cdot\|_2$ norm is equivalent to the $\|\cdot\|_2$ norm. Consequently, the vector space $L^2[a, b]$ equipped with the $_p\|\cdot\|_2$ norm is a Banach space, and any convergence results valid in $(L^2[a, b],_p \|\cdot\|_2)$ are still valid in $(L^2[a, b], \|\cdot\|_2)$ (and vice versa). If we define $\langle \cdot, \cdot \rangle_p$ by

$$\langle y_1, y_2 \rangle = \int_{[a,b]} p(x) y_1(x) y_2(x)\, dx,$$

then it is readily verified that $\langle \cdot, \cdot \rangle_p$ is an inner product and that $_p\|\cdot\|_2$ is the norm induced by this inner product. Thus the inner product space $(L^2[a, b], \langle \cdot, \cdot \rangle_p)$ is a Hilbert space. We use the (standard) notation $L^2([a, b], p)$ to denote this Hilbert space, with the abbreviation $L^2[a, b]$ for the space $L^2([a, b], 1)$.

If equation (9.5) satisfies the conditions of Theorem 9.4.1, the above arguments indicate that the normalized eigenfunctions

$\{y_n/_p\|y_n\|_2\}$ form an orthonormal set in the Hilbert space $L^2([a,b],p)$. In fact, this set forms an orthonormal basis for $L^2([a,b],p)$.

Theorem 9.4.2

Suppose that equation (9.5) satisfies the conditions of Theorem 9.4.1. Then the collection of normalized eigenfunctions forms an orthonormal basis for the Hilbert space $L^2([a,b],p)$.

Example 9-4-1: *Fourier Sine Series*

Consider the differential equation

$$y''(x) + \lambda y = 0, \tag{9.12}$$

with boundary conditions

$$y(0) = 0, \quad y(\pi) = 0. \tag{9.13}$$

If $\lambda < 0$, then the general solution to the differential equation (9.12) is

$$y(x) = Ae^{\sqrt{-\lambda}x} + Be^{-\sqrt{-\lambda}x},$$

where A and B are constants. The boundary conditions (9.13), however, indicate that $A = B = 0$, so that only the trivial solution is available in this case. Thus, this problem does not have any negative eigenvalues. If $\lambda = 0$, then the general solution is

$$Y(x) = Ax^2 + B,$$

where A and B are constants. Again, the boundary conditions imply that $A = B = 0$, so that $\lambda = 0$ cannot be an eigenvalue. If $\lambda > 0$, then the general solution is

$$y(x) = A\cos(\sqrt{\lambda}x) + B\sin(\sqrt{\lambda}x),$$

where A and B are constants. The condition $y(0) = 0$ implies that $A = 0$, and the condition $y(\pi) = 0$ implies that

$$B\sin(\sqrt{\lambda}\pi) = 0. \tag{9.14}$$

Equation (9.14) is satisfied for $B \neq 0$ only if $\lambda = n^2$ for some integer n, and in this case equation (9.12) has the nontrivial solution

$$y_n(x) = \sin(nx). \tag{9.15}$$

The set $\{n^2\}$ corresponds to the spectrum. Hence from Theorem 9.4.2 we know that the set $\{y_n/\|y_n\|_2\} = \{y_n\sqrt{2/\pi}\}$ forms an orthonormal basis for $L^2[0, \pi]$.

Example 9-4-2: *Mathieu Functions*
Consider the differential equation

$$y'' + (\lambda - 2\theta\cos(2x))y = 0, \tag{9.16}$$

along with the boundary conditions (9.13). Equation (9.16) is called *Mathieu's equation*, and θ is some fixed number. Here, $r(x) = p(x) = 1$ and $q(x) = -2\theta\cos(2x)$. Note that when $\theta = 0$ equation (9.16) reduces to equation (9.12). Now, unlike the previous example, we cannot solve equation (9.16) in closed form, and it is clear that the eigenvalues will depend on the parameter θ. Nonetheless, the above results indicate that for any θ, there is a set $\{\lambda_n(\theta)\}$ of eigenvalues with corresponding eigenfunctions that when normalized will yield an orthonormal basis for $L^2[0, \pi]$. The solutions to equation (9.16) are well-known special functions called (appropriately enough) *Mathieu functions*. The intricate details concerning these functions can be found in [28] or [41]. Suffice it here to say that corresponding to the spectrum $\{\lambda_n(\theta)\}$, Mathieu's equation has eigenfunctions $se_n(x, \theta)$ that are periodic with period 2π and reduce to sine functions[3] when $\theta = 0$. The Mathieu functions $\{se_n(x, \theta)\}$ thus form an orthogonal basis for $L^2[0, \pi]$.

The Sturm–Liouville problem can be posed under more general conditions. These generalizations lead to bases for L^2 that are widely used in applied mathematics and numerical analysis. The generalizations commonly made correspond to either:

(i) relaxing the conditions on p and r at the endpoints so that these functions may vanish (or even be discontinuous) at $x = a$ or $x = b$ (or both); or

(ii) posing the problem on an unbounded interval.

The general solutions to the differential equations with these modifications are usually unbounded on the interval, and the homo-

[3]The notation se_n comes from Whittaker and Watson [41] and denotes "sine-elliptic." There are "cosine-elliptic" functions ce_n with analogous properties.

geneous boundary conditions (9.6) are often replaced by conditions that ensure that the solution is bounded, or that limit the rate of growth of the function as x approaches a boundary point of the interval. These generalized versions of the Sturm–Liouville problem are called *singular Sturm–Liouville problems*. The theory underlying singular Sturm–Liouville problems and the corresponding results are more complicated than those for the regular Sturm–Liouville problems. For example, the spectrum may consist of isolated points or a continuum, and not every point in the spectrum need correspond to an eigenvalue.

The singular Sturm–Liouville problem is studied in some depth in [9] and [40]. More general references such as [5] and [11] give less detailed but clear, succinct accounts of the basic theory. We content ourselves here with a few examples that lead to well-known bases for L^2. The special functions arising in these examples have been studied in great detail by numerous authors, and we direct the reader to the aforementioned references as a starting point.

Example 9-4-3: *Legendre Polynomials*
Consider the *Legendre differential equation*

$$\frac{d}{dx}\left((1-x^2)\frac{dy}{dx}\right)+\lambda y=0 \tag{9.17}$$

on the interval $(-1,1)$. Note that $r(x)=1-x^2$ is zero at $x=\pm 1$, so that this equation leads to a singular Sturm–Liouville problem. The general solution to equation (9.17) can be found by using a power series method, which seeks solutions of the form

$$y(x)=\sum_{n=0}^{\infty}a_n x^n, \tag{9.18}$$

where the a_n's are constants (cf. [5] for details of the method). Substituting power series (9.18) into the Legendre differential equation and equating the coefficients of x^n to zero for $n=1,2,\ldots$ yields the recursive relation

$$a_{n+2}=\frac{n(n+1)-\lambda}{(n+1)(n+2)}.$$

Once a_0 and a_1 are specified, the above relation determines the other a_n's uniquely. Specifically, the recursive relation defines two linearly independent solutions y_e and y_o corresponding to the choices $a_0 = 1, a_1 = 0$ and $a_0 = 0, a_1 = 1$, respectively. Here y_e is an even solution and y_o is an odd solution. Suppose now that we require the solution to be *bounded* as $x \to \pm 1$. If $\lambda \neq n(n+1)$, we can apply the ratio test to the series defining y_e (or y_o) to establish that the radius of convergence is 1, and it can be shown that y_e (and y_o) are unbounded in the interval $(-1, 1)$. We thus need $\lambda = n(n+1)$ for some positive integer n to get bounded solutions. The eigenvalues for the problem are thus $\lambda_n = n(n+1)$. The corresponding eigenfunctions are the polynomials formed by the truncated series for y_e and y_o (modulo a scaling factor). Specifically, the eigenfunctions P_n corresponding to λ_n are defined by

$$P_n(x) = \sum_{m=0}^{M} (-1)^m \frac{(2n-2m)!}{2^n m!(n-m)!(n-2m)!} x^{n-2m},$$

where $M = n/2$, or $(n-1)/2$, whichever is an integer. For example,

$$P_0(x) = 1, \qquad P_1(x) = x,$$

$$P_2(x) = \tfrac{1}{2}(3x^2 - 1), \quad P_3(x) = \tfrac{1}{2}(5x^3 - 3x).$$

The peculiar form of the polynomial coefficients is standard: The polynomials have been scaled so that $P_n(1) = 1$ for all n. This corresponds to choosing the last coefficient c_n in the polynomial P_n to be

$$c_n = \frac{(2n)!}{2^n (n!)^2}$$

and using the recursive relation to get the lower-order coefficients. The functions P_n are called the *Legendre polynomials*. Note that the proof of orthogonality follows immediately from the self-adjointness of the operator. The boundary values for the P_k do not matter, because $r(1) = r(-1) = 0$. The Legendre polynomials, suitably normalized, form an orthonormal basis for $L^2[-1, 1]$.

Example 9-4-4: *Hermite Polynomials*
The Hermite differential equation is

$$\frac{d^2y}{dx^2} - 2x\frac{dy}{dx} + \lambda y = 0. \tag{9.19}$$

This equation is not in the self-adjoint form of equation (9.5), but since

$$\frac{d}{dx}\left(e^{-x^2}\frac{dy}{dx}\right) = e^{-x^2}\left(\frac{d^2y}{dx^2} - 2x\frac{dy}{dx}\right),$$

and $e^{-x^2} \neq 0$ for all $x \in \mathbb{R}$, equation (9.19) is equivalent to

$$\frac{d}{dx}\left(e^{-x^2}\frac{dy}{dx}\right) + \lambda e^{-x^2}y = 0. \tag{9.20}$$

The singular Sturm–Liouville problem consists in finding solutions y to equation (9.20) on the interval $(-\infty, \infty)$ such that $|y(x)|$ does not grow exponentially as $x \to \pm\infty$. Substituting the power series (9.18) into the equivalent equation (9.19) yields the recursive relation

$$a_{n+2} = \frac{2n - \lambda}{(n+1)(n+2)}a_n.$$

As with the Legendre equation, this recursive relation defines an even and an odd solution. If $\lambda \neq 2n$, then the ratio test indicates that the series converges for all $x \in \mathbb{R}$. The solutions in this case will have exponential growth. In order to meet the growth condition, we must therefore have that $\lambda = 2n$ for some $n = 0, 1, 2, \ldots$. If $\lambda = 2n$, then the series (9.18) reduces to a polynomial of degree n. The eigenfunctions are commonly given in the form

$$\begin{aligned} H_0(x) &= 1, \\ H_n(x) &= (-1)^n e^{x^2}\frac{d^n}{dx^n}\left(e^{-x^2}\right), \end{aligned}$$

and called *Hermite polynomials*. The first few Hermite polynomials are

$$H_1(x) = 2x, \qquad H_2(x) = 4x^2 - 2,$$

$$H_3(x) = 8x^3 - 12x, \quad H_4(x) = 16x^4 - 48x^2 + 12.$$

Note that for the self-adjoint equation (9.20), we have that $p(x) = e^{-x^2}$, and therefore for $m \neq n$,

$$\int_{(-\infty,\infty)} e^{-x^2}H_n(x)H_m(x)\,dx = 0.$$

It can be shown that the Hermite polynomials form an orthogonal basis for $L^2(\mathbb{R}, e^{-x^2})$. If e_n is defined by

$$e_n(x) = \frac{1}{\sqrt{2^n n! \sqrt{\pi}}} e^{-x^2/2} H_n(x),$$

then it can be shown that the set $\{e_n\}$ is an orthonormal basis for $L^2(\mathbb{R}, e^{-x^2})$.

Example 9-4-5: *Laguerre Polynomials*
A basis for $L^2(0, \infty)$ can be derived from the *Laguerre differential equation*

$$x\frac{d^2y}{dx^2} + (1-x)\frac{dy}{dx} + \lambda y = 0, \tag{9.21}$$

which in self-adjoint form is

$$\frac{d}{dx}\left(xe^{-x}\frac{dy}{dx}\right) + \lambda e^{-x} y = 0. \tag{9.22}$$

The singular Sturm–Liouville problem consists in solving equation (9.22) on the interval $(0, \infty)$ subject to the conditions $\lim_{x\to 0^+} y(x) < \infty$, $\lim_{x\to\infty} e^{-x} y(x) = 0$. As with the Hermite equation, the growth conditions lead to a restriction on the values of λ. For the Laguerre differential equation, the eigenvalues are $\lambda_n = n$ for $n = 1, 2, \ldots$. The standard representation for these eigenfunctions is

$$L_0(x) = 1,$$
$$L_n(x) = \frac{e^x}{n!}\frac{d^n}{dx^n}(x^n e^{-x}).$$

The functions L_n are polynomials of degree n; they are called *Laguerre polynomials*. The first few Laguerre polynomials are

$$L_1(x) = 1 - x, \qquad L_2(x) = 1 - 2x + \tfrac{1}{2}x^2,$$

$$L_3(x) = 1 - 3x + \tfrac{3}{2}x^2 - \tfrac{1}{6}x^3, \quad L_4(x) = 1 - 4x + 3x^2 - \tfrac{2}{3}x^3 - \tfrac{1}{24}x^4.$$

The Laguerre polynomials form an orthogonal basis for the space $L^2((0, \infty), e^{-x})$

The above examples are a small sample of the many "special functions" in mathematics that correspond to bases for L^2. The list of well-known special-function bases for L^2 is extensive. Aside from the references given above, the reader is also directed to the monograph of [19], which lists most the bases for L^2 in common use and gives another perspective on total sets for L^2.

Exercises 9-4:

1. Using the properties of the inner product and self-adjoint operators, prove that a Sturm–Liouville problem satisfying the conditions of Theorem 9.4.1 must have only real eigenvalues.

2. Solve equation (9.12) subject to the boundary conditions $y(0) = 0$, $y'(1) = 0$ and determine the eigenvalues and corresponding eigenfunctions.

3. Verify by direct calculation that the Hermite polynomials H_0, H_1, and H_2 of Example 9-4-4 form an orthogonal set in the space $L^2(\mathbb{R}, e^{-x^2})$.

9.5 Other Bases for L^2

The Sturm–Liouville problem can be used to generate numerous bases for L^2 spaces; however, any solution to equation (9.5) must be at least twice differentiable, and consequently, any basis derived from a Sturm–Liouville problem must consist of "smooth" functions with at least two derivatives. Given the nature of the functions in L^2, one expects that bases consisting of nonsmooth functions should also be available. In this section we present a brief example of such a basis for the space $L^2[0, 1]$.

The *Haar functions* $h_n : [0, 1] \to \mathbb{R}$ are defined by

$$h_1(x) = 1,$$
$$h_{2^k+\ell} = \begin{cases} 2^{k/2}, \text{ if } x \in \left[\frac{\ell-1}{2^k}, \frac{\ell-1/2}{2^k}\right), \\ -2^{k/2}, \text{ if } x \in \left[\frac{\ell-1/2}{2^k}, \frac{\ell}{2^k}\right], \\ 0, \text{ otherwise}, \end{cases}$$

where $\ell = 1, 2, \ldots, 2^k$ and $k = 0, 1, \ldots$. For example, if $n = 2$, then $k = 0$ and $\ell = 1$, so that

$$h_2(x) = \begin{cases} 1, & \text{if } x \in [0, \frac{1}{2}), \\ -1, & \text{if } x \in [\frac{1}{2}, 1]. \end{cases}$$

If $n = 3$, then $k = 1$ and $\ell = 1$, and thus

$$h_3(x) = \begin{cases} 2^{1/2}, & \text{if } x \in [0, \frac{1}{4}), \\ -2^{1/2}, & \text{if } x \in [\frac{1}{4}, \frac{1}{2}], \\ 0, & \text{if } x \in (\frac{1}{2}, 1]. \end{cases}$$

If $n = 4$, then $k = 1$, $\ell = 2$, and

$$h_4(x) = \begin{cases} 0, & \text{if } x \in [0, \frac{1}{2}), \\ 2^{1/2}, & \text{if } x \in [\frac{1}{2}, \frac{3}{4}], \\ -2^{1/2}, & \text{if } x \in (\frac{3}{4}, 1]. \end{cases}$$

The support of the Haar function $h_{2^k+\ell}$ is the interval $[(\ell - 1)/2^k, \ell/2^k]$. Figure 9.1 depicts the functions h_5 through h_8.

The set of Haar functions $H = \{h_n\}$ forms an orthonormal basis for $L^2[0, 1]$. The proof that H is a total set can be found in [19]. Here, we show that H is an orthonormal set in $L^2[0, 1]$. The normality of the Haar functions is simple to establish. Evidently, $\|h_1\|_2 = 1$; if $n = 2^k + \ell$, then

$$\begin{aligned} \|h_n\|_2^2 &= \int_{[0,1]} |h_n(x)|^2 \, dx \\ &= \int_{\ell-1/2^k}^{\ell/2^k} |h_n(x)|^2 \, dx \\ &= \int_{\ell-1/2^k}^{\ell/2^k} 2^k \, dx \\ &= 1. \end{aligned}$$

Hence $\|h_n\|_2 = 1$ for all $n = 1, 2, \ldots$. To establish orthogonality note first that $h_1 \perp h_n$ for all $n = 2, 3, \ldots$. Suppose that k is fixed, and

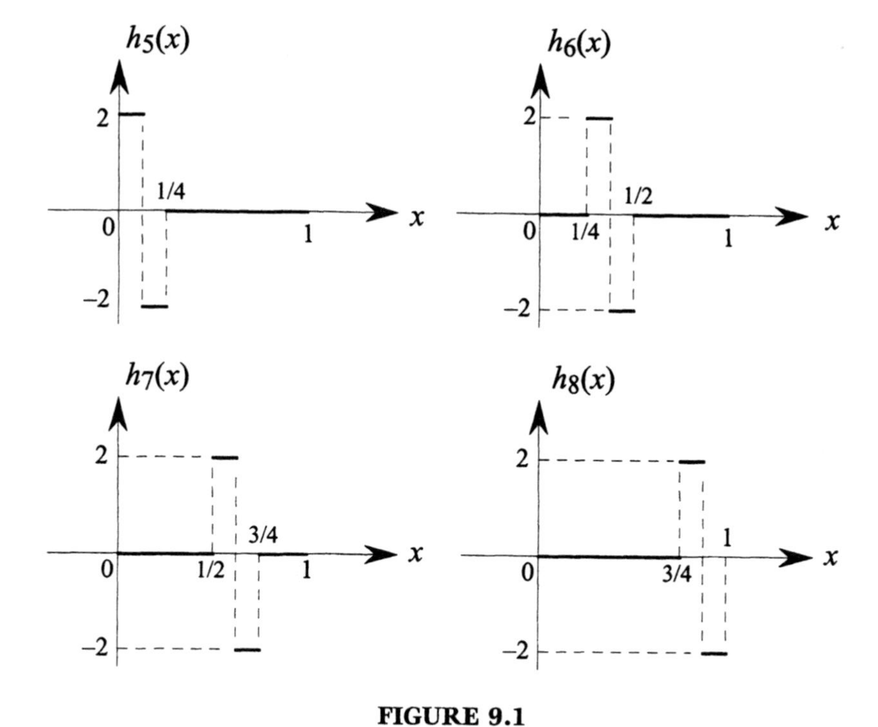

FIGURE 9.1

consider the functions $h_{2^k+\ell}$ for $\ell = 1, 2, \ldots, 2^k$. The support of $h_{2^k+\ell}$ is the interval $I_{k,\ell} = [(\ell-1)/2^k, \ell/2^k]$, and if $\ell_1 \neq \ell_2$, the set $I_{k,\ell_1} \cap I_{k,\ell_2}$ contains at most one point. Therefore, $h_{2^k+\ell_1}(x)\, h_{2^k+\ell_2}(x) = 0$ a.e., so that

$$\int_{[0,1]} h_{2^k+\ell_1}(x)\, h_{2^k+\ell_2}(x)\, dx = 0.$$

Consequently, we have that $h_{2^k+\ell_1} \perp h_{2^k+\ell_2}$ for all $\ell_1 \neq \ell_2$. Suppose now that $k_1 > k_2$ and consider the functions $h_{2^{k_1}+\ell_1}$, $h_{2^{k_2}+\ell_2}$ with supports I_{k_1,ℓ_1}, I_{k_2,ℓ_2}, respectively. Since $k_1 > k_2$, the length of I_{k_1,ℓ_1} is at most half that of I_{k_2,ℓ_2}, and since any endpoint in a support for a Haar function must be of the form $m/2^n$ for $m, n = 0, 1, 2 \ldots$, the set $I_{k_1,\ell_1} \cap I_{k_2,\ell_2}$ will be one of the following:

(i) the empty set;

(ii) a set consisting of a single endpoint of I_{k_1,ℓ_1};

(iii) the set I_{k_1,ℓ_1}.

Cases (i) and (ii) indicate that $h_{2^{k_1}+\ell_1}(x)h_{2^{k_2}+\ell_2}(x) = 0$ a.e. and therefore $h_{2^{k_1}+\ell_1} \perp h_{2^{k_2}+\ell_2}$. For case (c), note that in I_{k_1,ℓ_1}, the function $h_{2^{k_2}+\ell_2}$ does not change sign; hence,

$$\begin{aligned}\int_{[0,1]} h_{2^{k_1}+\ell_1}(x)h_{2^{k_2}+\ell_2}(x)\,dx &= \pm 2^{k_2/2}\int_{I_{k_1,\ell_1}} h_{2^{k_1}+\ell_1}(x)\,dx \\ &= 0.\end{aligned}$$

Thus, for case (iii), $h_{2^{k_1}+\ell_1} \perp h_{2^{k_2}+\ell_2}$. The above arguments indicate that $h_m \perp h_n$ for $m \neq n$, and the set H is consequently orthonormal.

Series of Haar functions may be used to represent measurable functions. If $f : [0,1] \to \mathbb{R}_e$ is finite a.e. on the interval $[0,1]$ and measurable, then there exists a series of the form $\sum_{n=1}^{\infty} \alpha_n h_n(x)$ that converges a.e. on the interval $[0,1]$ to f.

In passing we note that the Rademacher functions r_n of Exercises 9-2-3 do not form a basis for $L^2[0,1]$; however, another set of functions called the Walsh functions w_n can be formed from the products of Rademacher functions. The set of Walsh functions $\{w_n\}$ forms a basis for $L^2[0,1]$, and these functions have applications in probability and communication theory. For a short discussion of these functions the reader is referred to [19].

10 CHAPTER Epilogue

10.1 Generalizations of the Lebesgue Integral

The Lebesgue–Stieltjes integral may be generalized in any number of ways to accommodate, for example, higher dimensions. Rather than pursue extensions of this nature, we choose to describe informally and briefly a generalization of the Lebesgue integral whose origins go back to Newton.

Newton regarded the integral of a function as being the antiderivative of that function. Formally, a function $f : [a, b] \to \mathbb{R}$ is said to be *Newton integrable* on the interval $[a, b]$ if there exists a differentiable function F defined on $[a, b]$ such that $F'(x) = f(x)$ for all $x \in [a, b]$. The definition of the Newton integral is evidently very limiting: not even step functions, strictly speaking, have Newton integrals. On the other hand, there are functions that have Newton integrals but are not Lebesgue integrable. Consider, for example, the function F defined by

$$F(x) = \begin{cases} x^2 \sin(1/x^2) & \text{if } x \neq 0, \\ 0 & \text{if } x = 0. \end{cases}$$

Now, F is evidently differentiable at every point in the interval $(0, 1]$. At $x = 0$ the definition of the derivative can be used to establish that F is differentiable there as well. Consequently, the function f defined by $f(x) = F'(x)$ is Newton integrable in the interval $[0, 1]$. It can be shown, however, that this function is not Lebesgue integrable on $[0, 1]$. It is interesting to note that the improper Riemann integral also exists for f on the interval $(0, 1]$.

The essence of the problem with the Lebesgue integral is that the theory is restricted to absolutely convergent integrals. Recall that if a function f is Lebesgue integrable, then the function $|f|$ must also be Lebesgue integrable. This seems quite a harsh restriction, and it filters out conditionally convergent Riemann integrals as well as certain Newton integrals.

Newton's definition of an integral is very natural to the student of elementary calculus and useful, particularly in fields such as differential equations. The class of Newton integrable functions is not contained by the class of Lebesgue integrable functions, and this awkward situation was soon realized by mathematicians. Within fifteen years of Lebesgue's pioneering work two mathematicians arrived independently at a generalization of the Lebesgue integral that would mend the awkward gaps in the definition where improper Riemann integrals and/or Newton integrals exist, but the Lebesgue integral does not. In 1912, A. Denjoy made a generalization directly from the Lebesgue integral. The definition of the Denjoy integral proved a complicated affair, and as a result, some of its potential for applications and generalizations was lost. In 1914, O. Perron devised another integral that would also remedy the problem. Rather than start with the Lebesgue integral, Perron devised a definition based on upper and lower integrals defined by functions whose derivatives are respectively greater than and less than the given function. Roughly, a function A is said to be a *major function* of f on the interval $[a, b]$ if its derivative A' satisfies $A'(x) \geq f(x)$ for every $x \in [a, b]$. A function B is said to be a *minor function* for f if $-B$ is a major function for $-f$. A function f is said to be *Perron integrable* on the interval $[a, b]$ if f has both major and minor functions, and

$$-\infty < \inf\{A(b) - A(a)\} = \sup\{B(b) - B(a)\} < \infty.$$

Here, the infimum is taken over all major functions of f on $[a, b]$, and the supremum is taken over all minor functions of f on $[a, b]$. The common value of the infimum and supremum is defined to be the Perron integral of f on $[a, b]$. Note that Perron's extension immediately includes the class of Newton integrable functions. This is because the primitive F of a Newton integrable function f is at the same time a major and a minor function. Although the Perron integral patched up the gaps with the Lebesgue integral, it suffered from the same problems as the Denjoy integral.

The Perron integral and the Denjoy integral are defined very differently, and for a time it was thought that they characterized different functions as "integrable." As it turns out, however, these integrals are equal, i.e., f is Denjoy integrable if and only if it is Perron integrable. The integral is now commonly referred to as the Denjoy–Perron integral.

10.2 Riemann Strikes Back

The reader who has followed these chapters on the Lebesgue–Stieltjes integral is doubtless convinced of the vast superiority of the Lebesgue integral over the humble Riemann integral. Indeed, the Riemann integral is denigrated by many authors as merely a mathematical object of "historical" interest. At best, the integral is used as a pedagogical tool to introduce a "rigorous definition" of the integral in an elementary course in analysis. For example, in the 1930s Norbert Wiener [42] wrote in the introduction to his book on the Fourier integral,"However, the Riemann integral is of relatively little importance in the theory of Fourier series and integrals, save as the classical definition applying to continuous and 'step-wise continuous' functions." It is certainly true that the Lebesgue integral has won many resounding victories over the Riemann integral in fields such as Fourier analysis; it is also true that the Riemann integral suffers analytical deficiencies that are absent with the Lebesgue integral, but the victory is not complete, and good ideas (once accepted) are difficult to extinguish completely in mathematics. There is still an important realm where the Riemann integral reigns, *viz.*, conditionally convergent integrals. The (improper) Riemann integral does not suffer the same restrictions as the Lebesgue integral regarding absolute convergence, and this annoying fact perhaps motivated Lebesgue and some of his contemporaries (e.g., Denjoy and Perron) to search for a more all-embracing definition of an integral.

It is interesting to note that though many mathematicians have joined in the funeral chorus for the Riemann integral, the study of the Riemann integral has never really left the curriculum of mathematicians, owing to certain pedagogical advantages it has over the Lebesgue integral, and the need for conditionally convergent integrals in applications.

Riemann's approach to integration was as revolutionary as Lebesgue's; Riemann effectively divorced the integral from differentiation and brought the geometrical properties of the integral into focus. In the late 1950s, the Riemann approach was vindicated by R. Henstock and J. Kurzweil. Working separately, these mathematicians extended the Riemann integral to include Lebesgue integrable functions. Even more impressive, their Riemann-based definition of an integral also captured the more general Denjoy–Perron integral. The resulting integral is now called the *Henstock–Kurzweil integral*, and it has since been shown to be equivalent to the Denjoy–Perron integral,[1] i.e., f is Denjoy–Perron integrable if and only if it is Henstock–Kurzweil integrable. The key to their success in the extension was replacing Riemann's *uniformly* fine partitions of the integration interval with *locally* fine partitions. The use of locally fine partitions can be readily motivated by numerical examples. Consider, for example, the function f defined by $f(x) = x^{-1}\sin(1/x)$ in some interval $[\epsilon, 100]$, where ϵ is some small positive number. The graph of f shows that the function oscillates rapidly in the interval $[\epsilon, 1]$, but it then decreases steadily to 0 in the interval $[1, 100]$. If we wished to approximate efficiently and accurately the integral

[1]It is tempting now to refer to the common integral as the Denjoy–Perron–Henstock–Kurzweil integral, but one needs some patience in reading and typing such an appellation, and thus we eschew it. Often, if some proof requires the integral, then it is referred to as the Denjoy–Perron integral or the Henstock–Kurzweil integral, depending on which definition is to be used for the purposes of the proof.

of f over the interval $[\epsilon, 100]$, we would be inclined to use a much finer partition of the interval $[\epsilon, 1]$ than in the interval $[1, 100]$. Indeed, given some freedom, we would probably devise even more refined partitions for subintervals near the endpoint at $x = \epsilon$. The Henstock–Kurzweil approach allows the partition refinements in the integration interval to be nonuniform in the limit and thus "extra fine" where needed. Although the Henstock–Kurzweil integral is equivalent to the Denjoy–Perron integral, the Riemann-type definition of the former makes it a more tractable concept and thus more amenable to applications and generalizations.

In passing we note that these general integrals do not lead immediately to complete function spaces. The extension of the Lebesgue integral to include conditionally convergent integrals (the Denjoy–Perron integral) or the generalization of the Riemann integral to include the Lebesgue integral (the Henstock–Kurzweil integral) opens floodgates that neither Riemann nor Lebesgue can close without help.

10.3 Further Reading

Our approach to the Lebesgue–Stieltjes integral in this book has been pragmatic and arguably ostrich-like. Major results such as the monotone convergence theorem were stated without proof, and we seldom entertained generalizations or abstractions. The purist might rightfully claim that this approach is demeaning to the subject; however, there is little harm in viewing some of the cornucopia before a substantial investment in further study is made (if desired). Moreover, the Lebesgue–Stieltjes integral is no longer the exclusive preserve of mathematicians: It is an important tool in any subject that uses integrals. At any rate, there are numerous excellent texts that cover the Lebesgue–Stieltjes integral in depth resplendent with proofs and abstractions. Some elementary texts pitched approximately at the same level as this book include those by Pitt [30], [31], Priestley [32], and Weir [43], among others. These books develop the Lebesgue integral from the central concept of measure (which we did not emphasize). Pitt's books also cover applications of the Lebesgue–Stieltjes integral to geometry, harmonic analysis, and probability theory. Priestley's book is a particularly lively account of the integral[2] with many practical comments and examples.

There are several advanced accounts of integration theory available. Classic specialist references include Halmos [14] and Taylor [38], but many advanced analysis books such as Royden [36] and Rudin [37] also cover integration in depth. In addition, since the L^p spaces loom large in functional analysis, most books on this subject devote some time to the Lebesgue integral. Riesz and Nagy [35], for example, devote nearly a quarter of their book to the Lebesgue and Stieltjes integrals; other authors such as Hutson and Pym [22] and Yosida [44] give concentrated but general accounts of the theory. The reader is encouraged to explore these references as curiosity or the need for more refined details dictates.

Some of the reference cited above discuss the "post Lebesgue" integrals of Denjoy et al. These integrals have formed a nucleus of specialist literature somewhat apart from the normal texts on integration. Lee [27] and Pfeffer [29] provide a quite accessible reference on the Henstock–Kurzweil integral and its equivalence to the Denjoy–Perron integral. Gordon [13] also provides a basic self-contained account of these integrals along with the Lebesgue integral. The reader is also directed to the article by Bartle [4] for an introductory account.

Finally, the history of integration is interesting in its own right. Most accounts (such as ours) contain fleeting glimpses of the development of the integral. Hawkins [16] traces the history of the Riemann and Lebesgue integrals along with some early applications of the Lebesgue integral.

[2]Where else would you find yetis and nonmeasurable functions discussed in the same paragraph?

Appendix: Hints and Answers to Selected Exercises

Exercises 1-1

2. For example, $\sqrt{2}+(1-\sqrt{2}) = 1$ is rational, while $\sqrt{2}+\sqrt{2} = 2\sqrt{2}$ is irrational; $(\sqrt{2})(\sqrt{2}) = 2$ is rational, while $(\sqrt{2})(1+\sqrt{2}) = \sqrt{2}+2$ is irrational.

3. *Hint*: Can you be sure that b is an element of S^*?

4.

(a) *Hint*: Look at the proof that the set of all integers is countable.

(b) *Hint*: If the set of all irrational numbers were countable, what would that imply about the union of the set of all irrational numbers and the set of all rational numbers?

Exercises 1-3

1. Let lub and glb denote the least upper bound and the greatest lower bound respectively.

(a) lub $= 5$, glb $= 0$, both are in the set.

(b) lub $= 5$, glb $= 0$, 5 is in the set but 0 is not.

(c) lub $= +\infty$, glb $= -\infty$, neither is in the set.

(d) lub $= \frac{1}{2}$, glb $= 0$, neither is in the set.

(e) lub $= \sqrt{2}$, glb $= -\sqrt{2}$, neither is in the set.

(f) lub $= 4$, glb $= 3$, 4 is in the set but 3 is not.

(g) lub $= +\infty$, glb $= 0$, neither is in the set.

3. *Hint*: Show that $\sup S_2$ is an upper bound of S_1, and that $\inf S_2$ is a lower bound of S_1.

4. *Hint*: For (a), show that $c(\sup S)$ satisfies the definition of $\sup S^*$, in other words, show that $c(\sup S)$ is an upper bound of S^* and that $c(\sup S) \leq B$ for any upper bound B of S^*. Follow a similar strategy to show that $c(\inf S) = \inf S^*$, and also for part (b).

Exercises 2-1

1. *Hint*: To choose a sequence $\{a_n\}$ such that $a_n \uparrow M$, consider separately (i) M finite, (ii) $M = \infty$. For (i), choose $a_1 \in S$ such that $M - 1 < a_1 \leq M$ (using Theorem 1.3.2). Then choose $a_2 \in S$ such that $\max\{M - \frac{1}{2}, a_1\} \leq a_2 \leq M$, $a_3 \in S$ such that $\max\{M - \frac{1}{3}, a_2\} \leq a_3 \leq M$, and so on. For (ii), choose $a_1 \in S$ such that $a_1 > 1$. (Why is this always possible?) Then choose $a_2 \in S$ such that $a_2 > \max\{2, a_1\}$, $a_3 \in S$ such that $a_3 > \max\{3, a_2\}$, and so on. Use a similar approach to choose a sequence b_n such that $b_n \downarrow m$.

Exercises 2-4

3. *Hint*: Given $\epsilon > 0$, choose n such that $1/(n+1) < \epsilon$. It follows from the definition of f that $0 < x < 1/n \Rightarrow 0 < f(x) \leq 1/(n+1) < \epsilon$.

Exercises 2-6

1. *Hint*: Prove that $|f^+(x) - g^+(x)| \leq |f(x) - g(x)|$ for all $x \in I$ by considering the four cases $f(x) \geq 0$ and $g(x) \geq 0$, $f(x) \geq 0$ and $g(x) < 0$, $f(x) < 0$ and $g(x) \geq 0$, $f(x) < 0$ and $g(x) < 0$.

2. Similar to 1.

3. *Hint*: For each $x \in I$, $|f(x)| = |f(x) - g(x) + g(x)| \leq |f(x) - g(x)| + |g(x)|$, therefore $|f(x) - g(x)| \geq |f(x)| - |g(x)|$. Interchanging f and g gives $|f(x) - g(x)| \geq |g(x)| - |f(x)|$. Since $||f(x)| - |g(x)||$ must equal either $|f(x)| - |g(x)|$ or $|g(x)| - |f(x)|$, the result follows.

Exercises 2-7

1. *Hint*: Suppose f has bounded variation on I. Choose a point $a \in I$, and let x be any point in I. Denote by I_x the closed interval with endpoints a and x. Use the fact that $\{I_x\}$ is a partial subdivision of I, together with the definition of bounded variation, to obtain the required result.

2. *Hint*: For fg, use

$$\begin{aligned}\sum_{j=1}^{n} |f(b_j)g(b_j) - f(a_j)g(a_j)| &= \sum_{j=1}^{n} |f(b_j)g(b_j) - f(b_j)g(a_j) \\ &\quad + f(b_j)g(a_j) - f(a_j)g(a_j)| \\ &\le \sum_{j=1}^{n} |f(b_j)||g(b_j) - g(a_j)| \\ &\quad + \sum_{j=1}^{n} |g(a_j)||f(b_j) - f(a_j)|,\end{aligned}$$

and then use the result of Exercise 1 to obtain bounds for $|f(b_j)|$ and $|g(a_j)|$ that are independent of j.

3. *Hint*: In view of Exercise 1, for the first part you need prove only that if $\sup\{f(x) : x \in I\}$ and $\inf\{f(x) : x \in I\}$ are both finite (and f is monotone on I), then f has bounded variation on I. Using the notation for partial subdivisions introduced earlier, you can assume without loss of generality that

$$a_1 \le b_1 \le a_2 \le b_2 \le \cdots \le a_n \le b_n.$$

If f is monotone increasing on I, then

$$f(a_1) \le f(b_1) \le f(a_2) \le f(b_2) \le \cdots \le f(a_n) \le f(b_n),$$

and therefore for any partial subdivision of I we have

$$\begin{aligned}\sum_{j=1}^{n} |f(b_j) - f(a_j)| &= \sum_{j=1}^{n} (f(b_j) - f(a_j)) \\ &\le \sum_{j=1}^{n} (f(b_j) - f(a_j)) + \sum_{j=2}^{n} (f(a_j) - f(b_{j-1})) \\ &= f(b_n) - f(a_1) \\ &\le \sup\{f(x) : x \in I\} - \inf\{f(x) : x \in I\}.\end{aligned}$$

A similar argument shows that if f is monotone decreasing, then

$$\sum_{j=1}^{n} |f(b_j) - f(a_j)| \leq f(a_1) - f(b_n)$$
$$\leq \sup\{f(x) : x \in I\} - \inf\{f(x) : x \in I\}.$$

The first part then follows easily, and the second part is a straightforward application of the results already proved.

4. *Hint*: Use the construction described in the proof of Theorem 2.7.2.

5. **(a)** *Hint*: Use the fact that $|x \sin(1/x)| \leq |x|$ for all $x \neq 0$.

6. **(b)** *Hint*: If an interval I has finite endpoints a, b, then for any $x_1, x_2 \in I$ we have $|x_1^2 - x_2^2| = |x_1 + x_2||x_1 - x_2| \leq (2\max\{|a|, |b|\})|x_1 - x_2|$.

(c) *Hint*: Let $f(x) = x^2$, and assume that f is absolutely continuous on $I = (-\infty, \infty)$. Then by definition (choosing $\epsilon = 1$ in the definition), there exists a $\delta > 0$ such that $V_S(f, I) < 1$ for all partial subdivisions S of I for which the sum of the lengths of all the constituent intervals is less than δ. Considering in particular partial subdivisions consisting of a single interval $[n, n+\delta/2]$ (where n is a positive integer) leads to the desired contradiction.

Exercises 3-1

1. Suppose that $x \in [a, b]$. Then $x \in I_k$ for some $I_k \in P$ and $x \in I_{k'}$ for some $I_{k'} \in P'$. Now, P' is a refinement of P, so that $I_{k'} \subseteq I_k$. The result follows from the general inequalities $\inf\{f(x) : x \in I_{k'}\} \geq \inf\{f(x) : x \in I_k\}$ and $\sup\{f(x) : x \in I_{k'}\} \leq \sup\{f(x) : x \in I_k\}$.

2. Since the partition $Q = P \cup P'$ is a refinement of both P and P', Lemma 3.1.2 implies that $\overline{S}_P(f) \geq \overline{S}_Q(f)$ and $\underline{S}_Q(f) \geq \underline{S}_{P'}(f)$. Since $\overline{S}_Q(f) \geq \underline{S}_Q(f)$ for any partition Q, we have that

$$\overline{S}_P(f) \geq \overline{S}_Q(f) \geq \underline{S}_Q(f) \geq \underline{S}_{P'}(f),$$

and the lemma thus follows.

Exercises 4-1

1. **(b)** $\mu_\alpha((0,1)) = 1 - e^{-1}$, $\mu_\alpha([0,1]) = 3 - e^{-1}$, $\mu_\alpha((-1,1)) = 4 - e^{-1}$, $\mu_\alpha([0,0]) = 2$, $\mu_\alpha((-\infty,1)) = \infty$, $\mu_\alpha((0,\infty)) = 1$, $\mu_\alpha([0,\infty)) = 3$.

2. **(b)** $\mu_\alpha([-1,2)) = 4$, $\mu_\alpha((1,\infty)) = 2$, $\mu_\alpha((-\infty,4)) = 6$, $\mu_\alpha((0,2]) = 5$, $\mu_\alpha((\frac{1}{2},\frac{3}{2})) = 3$, $\mu_\alpha([1,3]) = 5$, $\mu_\alpha((1,3)) = 2$.

Exercises 4-2

1. $\alpha(x) = \begin{cases} 0, & \text{if } x < \lambda, \\ 1, & \text{if } x \geq \lambda. \end{cases}$

2. $\alpha(x) = \begin{cases} 0, & \text{if } x < \lambda_1, \\ i/n, & \text{if } \lambda_i \leq x < \lambda_{i+1} \ (i = 1, 2, \ldots, n-1), \\ 1, & \text{if } x \geq \lambda_n. \end{cases}$

Exercises 4-3

1. **(a)** $S \cup T = [1,8)$. $S \cap T = (2,3) \cup (4,5] \cup (6,7]$. $S - T = [1,2] \cup (5,6] \cup (7,8)$.

   (b) $S \cup T = [1,4] \cup [5,8)$. $S \cap T = (2,3) \cup [6,7]$. $S - T = [5,6)$.

   (c) $S \cup T = (1,4] \cup [5,7)$. $S \cap T = [2,2] \cup (5,6)$. $S - T = (1,2) \cup [5,5]$.

Exercises 4-4

1. $A_\alpha(\theta) = 0$.
2. $A_\alpha(\theta) = 0$.
3. $A_\alpha(\theta) = 3$.
4. $A_\alpha(\theta) = \frac{1}{2}$.
5. θ is not α-summable.

Exercises 4-5

2. *Hint*: The difficulty with this one is that it is too easy! Since $L_{\alpha^*}(|f|) = L_{\alpha^*}(f) = 0$, you can just take θ_n to be the zero function on $[0,1]$, for each $n = 1, 2, \ldots$.

Exercise 4-6

(a) Let n be the integer part of c, i.e., the largest integer not exceeding c. Then

$$\int_{[0,c]} f\,dx = \begin{cases} \frac{c-n}{n+1}, & \text{if } n \le c < n+\frac{1}{2}, \\ \frac{1+n-c}{n+1}, & \text{if } n+1/2 \le c < n+1. \end{cases}$$

(b) *Hint*: Following on part (a), show that

$$\left| \int_0^c f\,dx \right| \le \frac{1}{2(n+1)},$$

where n is the integer part of c.

(c) *Hint*: Show that if n is the integer part of c and $c \ge 1$, then

$$\int_0^c |f|\,dx = 1 + \frac{1}{2} + \frac{1}{3} + \cdots + \frac{1}{n} + \frac{c-n}{n+1}.$$

Exercises 5-1

2. *Hint*: To prove that $\max\{f(x), g(x)\} = f(x) + (g-f)^+(x)$ for all $x \in I$, consider separately the cases $f(x) \ge g(x)$ and $f(x) < g(x)$; similarly, for $\min\{f(x), g(x)\}$.

3. *Hint*: Use Theorem 5.1.5(ii) and Theorem 4.5.6.

4. *Hint*: Use Theorem 5.1.4 and Theorem 5.1.3.

Exercises 5-2

1. *Hint*: Use the same approach as was used in Example 4-5-2.

3. *Hint*: Use the fact that $g = 1-f$, where f is the function defined in Section 3.2, or use the fact that $g = 1$ a.e. to show that $\int_{[0,1]} g\,dx = 1$.

4. *Hint*: Define g^* on I by

$$g^*(x) = \begin{cases} g(x), & \text{if } g(x) \le f(x), \\ f(x), & \text{if } g(x) > f(x), \end{cases},$$

so that $g = g^*$ a.e. and $g^* \le f$ on I.

5. *Hint*: Show that $\chi_{S \cup T} \le \chi_S + \chi_T$ on $\mathbb{R}$.

Exercises 5-3

2. **(a)** $f = 0$.

Exercises 5-4

1. *Hint*: To obtain a sequence $\theta_1, \theta_2, \ldots$ of α-summable step functions on $\mathbb{R}$ such that $\lim_{n\to\infty} \theta_n = 1$ on $\mathbb{R}$, define

$$\theta_n(x) = \begin{cases} 1, & \text{if } -n \le x \le n, \\ 0, & \text{otherwise.} \end{cases}$$

2. **(b)** $\mu_\alpha([0,2)) = -1$, $\mu_\alpha([0,1]) = 0$, $\mu_\alpha([1,1]) = -1$, $\mu_\alpha((1,2)) = -1$.

   (c) 3.

Exercises 6-1

1. (i) $3(1-e^{-1})$, (ii) $4-3e^{-1}$, (iii) $4-3e^{-1}$, (iv) $4-3e^{-1}+e^{-2}$, (v) $8+e^{-2}$, (vi) 3.
2. **(b)** (i) 61, (ii) $e+e^2+e^3+e^4+2e^5$, (iii) 1, (iv) 2.
3. **(a)** $\frac{1}{2}(A+B)$.

   (b) $0, 1/n \sum_{i=1}^{n} \lambda_i$.

   (c) $\alpha(x) = \begin{cases} 0, & \text{if } x < \lambda_1, \\ \sum_{j=1}^{i} p_j, & \text{if } \lambda_i \le x < \lambda_{i+1} \ (i = 1, 2, \ldots, n-1), \\ 1, & \text{if } x \ge \lambda_n. \end{cases}$

   Mean $= \sum_{i=1}^{n} p_i \lambda_i$.

Exercises 6-3

1. $\lim_{t\to 0^+} \operatorname{erf}(t)/t = 2/\sqrt{\pi}$. $\lim_{t\to\infty} \operatorname{terfc}(t) = 0$.
5. $\frac{3}{t}\sin(t^3) - \frac{2}{t}\sin(t^2)$.

Exercise 7-4 Each repeated integral has the value $9\sin 1 + 4\sin 2$.

Exercises 8-1

1. **(a)** Properties (i), (iii), and (iv) are straightforward to establish. The real problem is showing that property (ii) is satisfied. Evidently, if $f = \mathbf{0}$, then $\|f\|_R = 0$. Suppose now that $\|f\|_R = 0$ but that $f \ne \mathbf{0}$. Since f is not identically zero on the interval $[a,b]$, there is some number $c \in [a,b]$ such that $|f(c)| > 0$. Since f is continuous on $[a,b]$ this means there is some interval $[\alpha, \beta] \subseteq [a,b]$ containing c such that $|f(x)| > 0$ for all

$x \in [\alpha, \beta]$. Now $\|f\|_R = \int_a^b |f(x)|\,dx \geq \int_\alpha^\beta |f(x)|\,dx > 0$, which contradicts the hypothesis that $\|f\|_R = 0$. Therefore, $\|\cdot\|_R$ satisfies property (ii).

(b) *Hint*: Let c be any number in the interval $[a, b]$ and let f be the function defined by $f(x) = 0$ if $x \neq c$ and $f(c) = 1$. What is the norm of this function?

3. *Hint*: Properties (i) and (ii) follow from the inequalities $\|f\|_{1,\infty} \geq \|f\|_\infty$ and $\|f\|_{1,1} \geq \|f\|_R$. Property (iv) follows from the inequality $\sup_{x\in[a,b]} |f(x) + g(x)| \leq \sup_{x\in[a,b]} |f(x)| + \sup_{x\in[a,b]} |g(x)|$.

4. *Hint*: $|S_{n+1} - S_n| \leq 1/10^n$, and if $n > m$, then $|S_n - S_m| = |(S_n - S_{n-1}) + (S_{n-1} - S_{n-2}) + \cdots + (S_{m+1} - S_m)|$.

5. Since $\|f\|_b \leq \beta\|f\|_a$ we can choose $\gamma = 1/\beta$. Similarly, we can choose $\delta = 1/\alpha$.

Exercises 8-2

1. Suppose that $a_n \to a$ as $n \to \infty$. Then, for any $\epsilon > 0$ there is an N such that $\|a_m - a\| < \epsilon/2$ whenever $m > N$. Now, $\|a_m - a\| = \|(a_m - a_n) + (a_n - a)\| \geq \|a_m - a_n\| - \|a_n - a\|$, and thus

$$\|a_m - a_n\| - \|a_n - a\| \leq \|a_m - a\| < \frac{\epsilon}{2},$$

so that if $n > N$, then

$$\|a_m - a_n\| - \frac{\epsilon}{2} < \frac{\epsilon}{2}.$$

Thus, for any $\epsilon > 0$ there is an N such that $\|a_m - a_n\| < \epsilon$ whenever $n, m > N$.

2. Let n be any positive integer. Note that $\sup_{x\in[-1,1]} |f_n(x) - f_{n+1}(x)|$ is achieved at $\hat{x} = 1/2^{n+1}$, where $f_{n+1}(\hat{x}) = 0$ and $f_n(\hat{x}) = 1 - 2^{m-(m+1)} = \frac{1}{2}$. Thus $\|f_n(x) - f_{n+1}(x)\|_\infty = \frac{1}{2}$ for all n, and $\{f_n\}$ cannot be a Cauchy sequence.

Exercises 8-3

1. (a) Let y be any element in Y and choose any $\epsilon > 0$. Since W is dense in Y, there is a $\tilde{w} \in W$ such that $\|\tilde{w} - y\| < \epsilon/2$. Similarly, since Z is dense in W, there is a $\tilde{z} \in Z$ such that

$\|\tilde{z} - \tilde{w}\| < \epsilon/2$. Consequently, for any $y \in Y$ and any $\epsilon > 0$ there is a $z \in Z$ such that $\|z - y\| < \epsilon$, i.e., Z is dense in Y.

(b) Use part (a) and the definition of completion.

2. *Hint*: Use the fact that the set of rational numbers is dense in the set of real numbers.

Exercises 8-5

1. **(a)** *Hint*: To show the triangle inequality ${}_r\|f+g\|_p \leq {}_r\|f\|_p + {}_r\|g\|_p$ apply the Minkowski inequality with $F = r^{1/p}f$ and $G = r^{1/p}g$.

(b) See the discussion after equation (9.11).

3. Since $f \in L^2[0,1]$, we have that $f \in L^1[0,1]$; since k is bounded, there is an $M < \infty$ such that $|k(x,y)| \leq M$ for all $(x,y) \in [0,1] \times [0,1]$. Hence,

$$|(Kf)(x)| \leq \int_{[0,1]} |k(x,\xi)||f(\xi)\, d\xi \leq M\|f\|_1,$$

and consequently

$$\|Kf\|_2^2 \leq \int_{[0,1]} M^2\|f\|_1^2\, d\xi = M^2\|f\|_1^2.$$

4. Part(iii): Suppose that $f \in L^1[a,b]$ and that $g \in L^\infty[a,b]$. Then

$$\begin{aligned}\int_{[a,b]} |f(x)g(x)|\, dx &\leq \|g\|_\infty \int_{[a,b]} |f(x)|\, dx \\ &= \|g\|_\infty \|f\|_1¥ < \infty,\end{aligned}$$

and therefore $fg \in L^1[a,b]$.
Part(iv): *Hint*: First establish the inequality

$$\int_{[a,b]} |f(x)|^p\, dx \leq \|f\|_\infty^p (b-a).$$

5. *Hint*: Note that

$$\left\{\int_{[a,b]} |f(x)|^p\, dx\right\}^{1/p} \leq \left\{\int_{[a,b]} \|f\|_\infty^p\, dx\right\}^{1/p}.$$

Exercises 9-1

1. Let $h = g/\|g\|$. Then

$$\begin{aligned} 0 &\le \|f - \langle f, h\rangle h\|^2 = \langle f - \langle f, h\rangle h, f - \langle f, h\rangle h\rangle \\ &= \|f\|^2 - \langle f, h\rangle\langle h, f\rangle - \overline{\langle f, h\rangle}\langle f, h\rangle + \langle f, h\rangle\overline{\langle f, h\rangle} \\ &= \|f\|^2 - |\langle f, h\rangle|^2; \end{aligned}$$

thus, $|\langle f, h\rangle| \le \|f\|$, and hence $|\langle f, g\rangle| \le \|f\|\,\|g\|$ for $g \neq 0$. (If $g = 0$, then the inequality follows immediately.)

3. *Hint*: Use the parallelogram equality, or verify by direct calculation.

Exercises 9-2

2. (a) For example,

$$\begin{aligned} \int_{-1}^{1} P_1(x)P_2(x)\,dx &= \int_{-1}^{1} x\frac{1}{2}(3x^2 - 1)\,dx \\ &= \frac{1}{4}\left[\frac{3}{2}x^4 - x^2\right]_{-1}^{1} = 0. \end{aligned}$$

(b) To normalize $P_0(x)$, note that $\|P_0\|^2 = \int_{-1}^{1} 1^2\,dx = 2$; thus, let $\phi_0(x) = \frac{1}{\sqrt{2}}P_0(x) = \frac{1}{\sqrt{2}}$. Similarly, $\|P_1\|^2 = \frac{3}{2}$, $\|P_2\|^2 = \frac{2}{5}$, so let $\phi_1(x) = \sqrt{\frac{2}{3}}P_1(x)$, $\phi_2(x) = \sqrt{\frac{5}{2}}P_2(x)$. The Fourier coefficients are given by $\langle e^x, \phi_n\rangle$ for $n = 0, 1, 2$. For example, $a_2 = \langle e^x, \phi_2\rangle = \int_{-1}^{1} e^x \frac{1}{2}(3x^2 - 1)\,dx = \sqrt{\frac{5}{2}}(e - \frac{7}{e})$.

4. Since $\mathcal{M}$ is a total orthonormal set, Parseval's formula is valid. Using the notation $a_k = \langle f, \phi_k\rangle$, $b_k = \langle g, \phi_k\rangle$ we thus have that $\|f\|^2 = \sum_{k=1}^{\infty} |a_k|^2$ and $\|g\|^2 = \sum_{k=1}^{\infty} |b_k|^2$. In addition, we also have that

$$\|f + g\|^2 = \sum_{k=1}^{\infty} |a_k + b_k|^2, \quad \|f + ig\|^2 = \sum_{k=1}^{\infty} |a_k + ib_k|^2.$$

Now,

$$\|f + g\|^2 = \|f\|^2 + 2\text{Re}\,\langle f, g\rangle + \|g\|^2,$$

so that

$$\begin{aligned}\|f\|^2 + 2\mathrm{Re}\,\langle f,g\rangle + \|g\|^2 &= \sum_{k=1}^{\infty} |a_k + b_k|^2 \\ &= \sum_{k=1}^{\infty} |a_k|^2 + 2\sum_{k=1}^{\infty} \mathrm{Re}\,(a_k\overline{b_k}) + \sum_{k=1}^{\infty} |b_k|^2 \\ &= \|f\|^2 + 2\sum_{k=1}^{\infty} \mathrm{Re}\,(a_k\overline{b_k}) + \|g\|^2;\end{aligned}$$

consequently, $\mathrm{Re}\,\langle f,g\rangle = \sum_{k=1}^{\infty} \mathrm{Re}\,(a_k\overline{b_k})$. A similar argument applied to $\|f + ig\|$ indicates that $\mathrm{Im}\,\langle f,g\rangle = \sum_{k=1}^{\infty} \mathrm{Im}\,(a_k\overline{b_k})$, and hence the result follows.

Exercises 9-3

1. **(a)** $P_{\mathcal{M}}x = 2(\sin(x) - \frac{1}{2}\sin(2x) + \frac{1}{3}\sin(3x) - \frac{1}{4}\sin(4x) + \cdots$.

2. **(a)** $P_{\mathcal{M}}g(x) = \sum_{-\infty}^{\infty} c_n e^{2n\pi i x}$, where

$$c_n = \begin{cases} \frac{-2i}{n\pi}, & \text{if } n \text{ is odd,} \\ 0, & \text{if } n \text{ is even.} \end{cases}$$

(b) $\|g\|^2 = \int_{-1/2}^{1/2} g^2(x)\,dx = 1 = \sum_{-\infty}^{\infty} |c_n|^2 = 2\sum_{k=1}^{\infty}\left(\frac{1}{(2k+1)\pi}\right)^2$.

3. Since f' is continuous on the interval $[-\pi,\pi]$, there exist numbers M and M' such that $|f(x)| \le M$ and $|f'(x)| \le M'$ for all $x \in [-\pi,\pi]$. Now,

$$\int_{-\pi}^{\pi} f(x)\cos(\lambda_n x)\,dx = \frac{1}{\lambda_n}\left[\sin(\lambda_n x)\right]_{-\pi}^{\pi} - \int_{-\pi}^{\pi} \frac{1}{\lambda_n} f'(x)\sin(\lambda_n x)\,dx,$$

and therefore

$$\begin{aligned}&\left|\int_{-\pi}^{\pi} f(x)\cos(\lambda_n x)\,dx\right| \\ &\quad\le \frac{1}{|\lambda_n|}\left(\left|\left[f(x)\sin(\lambda_n x)\right]_{-\pi}^{\pi}\right| + \int_{-\pi}^{\pi} \left|f'(x)\sin(\lambda_n x)\right|\,dx\right) \\ &\quad\le \frac{1}{|\lambda_n|}\left(2M + 2\pi M'\right).\end{aligned}$$

Since $|\lambda_n| \to \infty$ as $n \to \infty$, $\int_{-\pi}^{\pi} f(x)\cos(\lambda_n x)\,dx \to 0$ as $n \to \infty$. The limit for the sine integral can be established using the same arguments.

Exercises 9-4

1. Suppose λ is an eigenvalue for the operator $\mathcal{L}$. Then $\mathcal{L}y = -\lambda py$, and $\langle \mathcal{L}y, y\rangle = \langle -\lambda py, y\rangle$. If $\mathcal{L}$ is self-adjoint, then $\langle \mathcal{L}y, y\rangle = \langle y, \mathcal{L}y\rangle$; thus,

$$\langle \mathcal{L}y, y\rangle = -\lambda\langle py, y\rangle = \langle y, \mathcal{L}y\rangle = \langle y, -\lambda py\rangle = -\overline{\lambda}\langle \overline{p}y, y\rangle.$$

Since p is a real-valued function, we must have that $\lambda = \overline{\lambda}$, i.e, λ is real.

2. The eigenvalues are $\lambda_n = \left(\frac{(2n+1)\pi}{2}\right)^2$ for $n = 0, 1, 2\ldots$; corresponding eigenfunctions are $\phi_n = \sin\left(\frac{(2n+1)\pi}{2}x\right)$.

References

[1] Adams, R.A., *Sobolev Spaces*, Academic Press, 1975.

[2] Ahlfors, L., *Complex Analysis*, 2nd edition, McGraw-Hill Book Co., 1966.

[3] Arfken, G., *Mathematical Methods for Physicists*, 2nd edition, Academic Press, 1970.

[4] Bartle, R.G., "A return to the Riemann integral", *Amer. Math. Monthly*, **103** (1996) pp. 625–632

[5] Birkhoff, G. and Rota, G., *Ordinary Differential Equations*, 4th edition, John Wiley and Sons, 1989.

[6] Bromwich, T.A., *An Introduction to the Theory of Infinite Series*, Macmillan and Co., 1926.

[7] Carleson, L. "Convergence and growth of partial sums of Fourier Series" *Acta Math.*, **116**, (1966) pp. 135–157

[8] Churchill, R.V., *Fourier Series and Boundary Value Problems*, 2nd edition, McGraw-Hill Book Co., 1963.

[9] Coddington, E.A. and Levinson, N., *Theory of Ordinary Differential Equations*, McGraw-Hill Book Co., 1955.

[10] Conway, J.B., *Functions of One Complex Variable I*, 2nd edition, Springer-Verlag, 1978.

[11] Courant, R. and Hilbert, D., *Methods of Mathematical Physics*, volume 1, John Wiley and Sons, 1953.

[12] Dunford, N. and Schwartz, J.T., *Linear Operators*, parts I, II, III, John Wiley and Sons, 1971.

[13] Gordon, R.A., *The Integrals of Lebesgue, Denjoy, Perron, and Henstock*, American Math Soc., 1994

[14] Halmos, P.R., *Measure Theory*, Springer-Verlag, 1974.

[15] Hardy, G.H. and Rogosinski, W.W., *Fourier Series*, 3rd edition, Cambridge University Press, 1956

[16] Hawkins, T., *Lebesgue's Theory of Integration, Its Origins and Development*, The University of Wisconsin Press, 1970.

[17] Heider, L.J. and Simpson, J.E., *Theoretical Analysis*, W.B. Saunders Co., 1967.

[18] Hewitt, E. and Stromberg, K., *Real and Abstract Analysis*, Springer-Verlag, 1969.

[19] Higgins, J.R., *Completeness and Basis Properties of Sets of Special Functions*, Cambridge University Press, 1977.

[20] Hille, E., *Analytic Function Theory*, volume II, Ginn and Co., 1959.

[21] Hoffman, K., *Banach Spaces of Analytic Functions*, Prentice-Hall, 1962.

[22] Hutson, V. and Pym, J.S., *Applications of Functional Analysis and Operator Theory*, Academic Press, 1980.

[23] Ince, E.L., *Ordinary Differential Equations*, Longmans, Green and Co., 1927.

[24] Körner, T.W., *Fourier Analysis*, Cambridge University Press, 1988.

[25] Kreyszig, E., *Introductory Functional Analysis with Applications*, John Wiley and Sons, 1978.

[26] Kreyszig, E., *Advanced Engineering Mathematics*, 4th edition, John Wiley and Sons, 1979.

[27] Lee, P., *Lanzhou Lectures on Henstock Integration*, World Scientific, 1989.

[28] McLachlan, N.W., *Theory and Application of Mathieu Functions*, Oxford University Press, 1947.

[29] Pfeffer, W.F., *the Riemann Approach to Integration: Local Geometric Theory*, Cambridge University Press, 1993.

[30] Pitt, H.R., *Integration, Measure and Probability*, Oliver and Boyd, 1963.

[31] Pitt, H.R., *Measure and Integration for Use*, Oxford University Press, 1985.

[32] Priestley, H.A., *Introduction to Integration*, Oxford University Press, 1997.

[33] Pryce, J.D. *Basic Methods of Linear Functional Analysis*, Hutchinson and Co., 1973.

[34] Richtmyer, R.D. *Principles of Advanced Mathematical Physics*, volume I, Springer-Verlag, 1978.

[35] Riesz, F. and Nagy, B., *Functional Analysis*, Frederick Ungar Publishing Co., 1955.

[36] Royden, H.L., *Real Analysis*, Macmillan and Co., 1963.

[37] Rudin, W., *Real and Complex Analysis*, 2nd edition, McGraw-Hill Book Co., 1974.

[38] Taylor, A.E., *General Theory of Functions and Integration*, Blaisdell Publishing Co., 1965.

[39] Titchmarsh, E.C., *The Theory of Functions*, 2nd edition, Oxford University Press, 1939.

[40] Titchmarsh, E.C., *Eigenfunction Expansions*, Part I, 2nd edition, Oxford University Press, 1962.

[41] Whittaker, E.T. and Watson, G.N., *A Course of Modern Analysis*, 4th edition, Cambridge University Press, 1952.

[42] Wiener, N., *The Fourier Integral and Certain of Its Applications*, Cambridge University Press, 1933.

[43] Weir, A.J., *Lebesgue Integration and Measure*, Cambridge University Press, 1973.

[44] Yosida, K. *Functional Analysis*, 6th edition, Springer-Verlag, 1980.

[45] Zygmund, A., *Trigonometric Series*, volumes I and II, Cambridge University Press, 1959.

Index

Undergraduate Texts in Mathematics

(continued from page ii)

Hilton/Holton/Pedersen: Mathematical Reflections: In a Room with Many Mirrors.
Iooss/Joseph: Elementary Stability and Bifurcation Theory. Second edition.
Isaac: The Pleasures of Probability. *Readings in Mathematics.*
James: Topological and Uniform Spaces.
Jänich: Linear Algebra.
Jänich: Topology.
Kemeny/Snell: Finite Markov Chains.
Kinsey: Topology of Surfaces.
Klambauer: Aspects of Calculus.
Lang: A First Course in Calculus. Fifth edition.
Lang: Calculus of Several Variables. Third edition.
Lang: Introduction to Linear Algebra. Second edition.
Lang: Linear Algebra. Third edition.
Lang: Undergraduate Algebra. Second edition.
Lang: Undergraduate Analysis.
Lax/Burstein/Lax: Calculus with Applications and Computing. Volume 1.
LeCuyer: College Mathematics with APL.
Lidl/Pilz: Applied Abstract Algebra. Second edition.
Logan: Applied Partial Differential Equations.
Macki-Strauss: Introduction to Optimal Control Theory.
Malitz: Introduction to Mathematical Logic.
Marsden/Weinstein: Calculus I, II, III. Second edition.
Martin: The Foundations of Geometry and the Non-Euclidean Plane.
Martin: Geometric Constructions.
Martin: Transformation Geometry: An Introduction to Symmetry.
Millman/Parker: Geometry: A Metric Approach with Models. Second edition.
Moschovakis: Notes on Set Theory.
Owen: A First Course in the Mathematical Foundations of Thermodynamics.
Palka: An Introduction to Complex Function Theory.
Pedrick: A First Course in Analysis.
Peressini/Sullivan/Uhl: The Mathematics of Nonlinear Programming.
Prenowitz/Jantosciak: Join Geometries.
Priestley: Calculus: A Liberal Art. Second edition.
Protter/Morrey: A First Course in Real Analysis. Second edition.
Protter/Morrey: Intermediate Calculus. Second edition.
Roman: An Introduction to Coding and Information Theory.
Ross: Elementary Analysis: The Theory of Calculus.
Samuel: Projective Geometry. *Readings in Mathematics.*
Scharlau/Opolka: From Fermat to Minkowski.
Schiff: The Laplace Transform: Theory and Applications.
Sethuraman: Rings, Fields, and Vector Spaces: An Approach to Geometric Constructability.
Sigler: Algebra.
Silverman/Tate: Rational Points on Elliptic Curves.
Simmonds: A Brief on Tensor Analysis. Second edition.
Singer: Geometry: Plane and Fancy.
Singer/Thorpe: Lecture Notes on Elementary Topology and Geometry.
Smith: Linear Algebra. Third edition.
Smith: Primer of Modern Analysis. Second edition.
Stanton/White: Constructive Combinatorics.
Stillwell: Elements of Algebra: Geometry, Numbers, Equations.
Stillwell: Mathematics and Its History.
Stillwell: Numbers and Geometry. *Readings in Mathematics.*
Strayer: Linear Programming and Its Applications.

Thorpe: Elementary Topics in Differential Geometry.

Toth: Glimpses of Algebra and Geometry. *Readings in Mathematics.*

Troutman: Variational Calculus and Optimal Control. Second edition.

Valenza: Linear Algebra: An Introduction to Abstract Mathematics.

Whyburn/Duda: Dynamic Topology.

Wilson: Much Ado About Calculus.

Printed in the United States
122617LV00005BG/5/A

9 780387 950129